Wireless Sensor and Ad Hoc Networks Under Diversified Network Scenarios

For a complete listing of titles in the
Artech House Mobile Communications Library,
turn to the back of this book.

Wireless Sensor and Ad Hoc Networks Under Diversified Network Scenarios

Subir Kumar Sarkar

ARTECH HOUSE
BOSTON | LONDON
artechhouse.com

Library of Congress Cataloging-in-Publication Data
A catalog record for this book is available from the U.S. Library of Congress.

British Library Cataloguing in Publication Data
A catalogue record for this book is available from the British Library.

Cover design by Vicki Kane

ISBN 13: 978-1-60807-468-6

© 2012 ARTECH HOUSE
685 Canton Street
Norwood, MA 02062

All rights reserved. Printed and bound in the United States of America. No part of this book may be reproduced or utilized in any form or by any means, electronic or mechanical, including photocopying, recording, or by any information storage and retrieval system, without permission in writing from the publisher.

All terms mentioned in this book that are known to be trademarks or service marks have been appropriately capitalized. Artech House cannot attest to the accuracy of this information. Use of a term in this book should not be regarded as affecting the validity of any trademark or service mark.

10 9 8 7 6 5 4 3 2 1

Dedicated to my beloved parents,
Late Girish Chandra Sarkar
Late Prafullanalini Sarkar,
&
my in-laws,
Late Sudhir Kumar Bose
Late Santa Bose

Contents

	Preface	xv
1	**Introduction**	**1**
1.1	Overview of Wireless Networks	1
1.2	Types of Wireless Communication	3
1.2.1	Wireless Personal Area Networks Bluetooth	3
1.2.2	Wireless Local Area Networks 802.11	3
1.2.3	Wireless Metropolitan Area Networks WiMAX	3
1.2.4	Wireless Wide Area Networks	5
1.3	The IEEE 802.11 Family	5
1.4	Cellular Network	7
1.4.1	Existence of Mobile Generation and 4G	7
1.4.2	WiMAX	9
1.5	Mobile Ad Hoc Networks	11
1.6	Wireless Sensor Networks	11
1.7	Bluetooth	13
1.8	Organization of the Book	14
	References	16

2	**Fundamentals of Wireless Mobile Ad Hoc Networks**	23
2.1	Introduction	23
2.2	Literature Background	24
2.3	Applications of Mobile Ad Hoc Networks	26
2.3.1	Characteristics of MANETs	27
2.4	Medium Access Control Layer	28
2.5	Topology Control	29
2.6	Routing Protocols	30
2.7	Broadcasting	31
2.8	Multicasting	33
2.9	Internet Connectivity for Mobile Ad Hoc Networks	34
2.10	Security in Mobile Ad Hoc Networks	35
	References	36
3	**Scenario Based Performance Analysis of Various Routing Protocols in MANETs**	41
3.1	Introduction	41
3.2	Literature Background	43
3.3	Properties Desired in a Routing Protocol	45
3.4	Discussion of Various Routing Protocols	46
3.4.1	Ad Hoc On Demand Distance Vector Routing Protocol	47
3.4.2	Ad Hoc On Demand Distance Vector Routing Algorithm by Uppsala University (AODVUU)	48
3.4.3	Ad Hoc On Demand Multipath Distance Vector Routing Algorithm	49
3.4.4	Reverse Ad Hoc On Demand Distance Vector Routing Algorithm (RAODV)	49
3.4.5	Dynamic Source Routing Algorithm (DSR)	50
3.4.6	Destination Sequence Distance Vector Routing Algorithm (DSDV)	51

3.4.7	Dynamic MANET On Demand (DYMO) Routing Protocol	52
3.4.8	Fisheye State Routing Protocol (FSR)	54
3.4.9	Location-Aided Routing (LAR) Protocol	55
3.4.10	Optimized Link State Routing Protocol	58
3.4.11	Temporally Ordered Routing Algorithm (TORA)	59
3.5	Network Simulator, NS-2	60
3.6	Simulation Environment	61
3.7	Result Analysis	63
3.7.1	Evaluation of Performance Metrics	63
	References	83

4 An Empirical Study of Various Mobility Models in MANETs — 87

4.1	Introduction	87
4.2	Literature Background	89
4.3	Description of Various Mobility Models	90
4.3.1	Random Waypoint Mobility Model	90
4.3.2	Random Waypoint Mobility Model with Steady State	90
4.3.3	Random Walk Mobility Model with Reflection	92
4.3.4	Random Walk Mobility Model with Wrapping	93
4.3.5	Gauss-Markov Mobility Model	93
4.3.6	Reference Point Group Mobility Model	94
4.3.7	Group Force Mobility Model	94
4.3.8	Manhattan Mobility Model	96
4.3.9	Levy Walk Mobility Model	96
4.3.10	Community Based Mobility Model	97
4.3.11	Semi-Markov Smooth Mobility Model	98
4.4	Simulation Environment	99
4.5	Result Analysis	100
4.5.1	Evaluation of Performance Metrics with Mobility Speed	100
4.5.2	Evaluation of Performance Metrics for Different Network Scenarios	103
	References	107

5	**Overhead Control Mechanism and Analysis in MANET**	**111**
5.1	Introduction	111
5.2	Literature Background	113
5.3	Overhead Analysis in Hierarchical Routing Scheme	114
5.4	Overhead Minimizing Techniques and Analysis Using Clustering Mechanisms	116
5.5	Overhead Minimization by Header Compression	120
5.6	Overhead Minimization for Ad Hoc Networks Connected to the Internet	121
5.7	Energy Models	123
5.8	Simulation	124
5.9	Result Analysis	125
5.9.1	Performance Evaluation of Different Energy Models	125
5.9.2	Performance Evaluation of Energy Consumed with Mobility Speed	128
	References	130
6	**Study of Various Issues in WSNs and Analysis of Wireless Sensor and Actor Network Scenarios**	**133**
6.1	Introduction	133
6.2	Literature Background	135
6.3	Applications of WSNs	135
6.4	Hardware and Software Issues in WSN	136
6.4.1	Hardware Issues	136
6.4.2	Software Issues	138
6.5	Issues in Radio Communication	139
6.6	Design Issues of MAC Protocols	141
6.6.1	Classification of MAC Protocols	142
6.7	Deployment	144

6.8	Localization	145
6.9	Synchronization	146
6.10	Calibration	147
6.11	Network Layer Issues	148
6.11.1	Classification of Routing Protocols in WSNs	149
6.11.2	Popular Routing Protocols	150
6.12	Transport Layer Issues	152
6.13	Data Aggregation and Dissemination	154
6.14	Database Centric and Querying	156
6.15	Programming Models for Sensor Networks	160
6.16	Middleware	161
6.17	Quality of Service	162
6.18	Security	164
6.19	Wireless Sensor and Actor Network Architecture	165
6.20	Simulation Environment	166
6.21	Result Analysis	167
6.21.1	Packet Delivery Ratio Based Performance Evaluation	167
6.21.2	Average End-to-End Delay Based Performance Evaluation	169
6.21.3	Control Packet Overhead Based Performance Evaluation	170
6.21.4	Throughput Based Performance Evaluation	172
	References	173
7	**Performance Analysis of IEEE 802.15.4 with Sink Nodes in WSN Scenario**	**181**
7.1	Introduction	181
7.2	Literature Background	182
7.3	Overview of IEEE 802.15.4 and Its Characteristics	184
7.3.1	Network Topologies	185

7.3.2	LR-WPAN Protocol Architecture	186
7.3.3	The Superframe Structure	187
7.3.4	Data Transmission	188
7.3.5	Slotted CSMA-CA Mechanism (Beacon Enabled)	189
7.3.6	Starting and Maintaining PANs	192
7.3.7	Association and Disassociation	193
7.3.8	Synchronization	194
7.4	Data Gathering Paradigm	194
7.5	Simulation Environment	196
7.6	Result Analysis	197
7.6.1	Evaluation of Various Metrics with Traffic Load	198
7.6.2	Evaluation of Various Metrics with Simulation Time	200
7.6.3	Evaluation of Performance Metrics with Different Network Scenarios	201
	References	208

8 Performance Evaluation and Traffic Load Effect on Patrimonial ZigBee Routing Protocols in WSNs — 211

8.1	Introduction	211
8.2	Literature Background	212
8.3	Patrimonial ZigBee Routing Protocols	214
8.4	Traffic Generators	214
8.4.1	Exponential On-Off Traffic	214
8.4.2	Pareto On-Off Traffic	215
8.4.3	CBR On-Off Traffic	215
8.4.4	Poisson On-Off Traffic	215
8.5	Traffic Model	216
8.6	Simulation Environment	216
8.7	Result Analysis	218
8.7.1	Performance Evaluation of Various Protocols Under Network Scenarios	218
	References	229

9	**Applications and Recent Developments**	**231**
9.1	Introduction	231
9.2	Applications and Opportunities	232
9.3	Typical Applications	233
9.3.1	Academic Environment	234
9.3.2	Industrial or Corporate Environment	234
9.3.3	Health Care	234
9.3.4	Defense (Army, Navy, and Air Force)	235
9.3.5	Disaster Situation for Search and Rescue Operations	236
9.3.6	Traffic Management and Monitoring	236
9.3.7	Other Applications	236
9.4	Primary Issues for Wireless Networks	237
9.5	Recent Developments	238
9.6	Active Research Areas	239
9.7	Challenges and Future Scope	241
9.7.1	Research Challenges	242
	References	244
	About the Author	**249**
	Index	**251**

Preface

Mobile communications and wireless networking technologies have seen a thriving development and a steep growth in research in recent times because of their significant features. Driven by technological advancements as well as application demands, various classes of communication networks have emerged, such as ad hoc networks, cellular networks, sensor networks, and mesh networks.

Ad hoc networks are autonomous systems that comprise a collection of mobile nodes that use wireless transmission for communication. These networks are self-controlled, self-organized, and self-configured infrastructure-less networks. They can be developed anywhere and anytime because of their very simple infrastructure setup and minimal central administration. If any two nodes are out of range, then connectivity is established by hopping through various intermediate nodes. But if any two nodes are within the transmission range of each other, then the connectivity is established in a peer-to-peer manner.

Wireless sensor networks are special type of ad hoc network, used to provide a wireless communication infrastructure that allows us to instrument, observe, and respond to phenomena in the natural environment and in our physical and cyber infrastructure. Routing in wireless sensor networks like mobile ad hoc networks is very challenging due to the inherent characteristics that distinguish these networks from other wireless networks, like mobile ad hoc networks or cellular networks.

A large number of sensor devices are deployed in the field to create a sensor network for both monitoring and control purposes. However, various limits like dependability of the network, energy supply, cost, maintenance, and reliability of operation persist. This book depicts comprehensive resources on the recent ideas and simulation results in the area of mobile ad hoc networks and sensor networks explored by leading experts in the area, from both industry and academia, with diversified network scenarios. I hope that the unifying theme throughout the book will expose industry researchers interested in sensor network systems and applications, as well as students/researchers/academics who wish to pursue research in this area, to the many exciting and open research problems still present and prepare them in this nascent area for new development. This book is the outcome of my teaching and research experiences in the Department of Electronics and Telecommunication Engineering, Jadavpur University, India.

Coverage of the book includes the following:

1. Introduction
2. Fundamentals of Wireless Mobile Ad Hoc Networks
3. Scenario-Based Performance Analysis of Various Routing Protocols in MANETs
4. An Empirical Study of Various Mobility Models in MANETs
5. Overhead Control Mechanism and Analysis in MANETs
6. Study of Various Issues in WSNs and Analysis of Wireless Sensor and Actor Network Scenario
7. IEEE 802.15.4 Standards and Its Performance
8. Performance Evaluation and Traffic Load Effect on Patrimonial ZigBee Routing Protocols in WSNs
9. Applications and Recent Developments

I wish to thank many individuals, without whose contributions the present book would not have been possible. I would like to express my deep sense of gratitude to the reviewer—the reviewer's numerous suggestions and critical remarks have contributed immensely to the manuscript. Parts of the book are based on research results obtained by my Ph.D. scholars, Dr. T. G. Basavaraju and Dr. S. Gowrishankar. I heartily acknowledge their untiring support. I wish to thank my other research scholars, especially Sri Rajanna KM for his untiring effort in typing the various versions of the manuscript and in making continuous communication with the editor and production editor at Artech House to finalize the manuscript. I also thank my other research scholars and students: Sri. Abhik Roy, Sri. Guruprasad Mishra, Sri. Khondrom Joson Singh, Sri. Sanjoy Deb, Sri. Saptarsi Ghosh, Sri. Anindya Jana, Sri. Bijoy Kantho, Dr. S. R. Biradar, Sri. Koushik Majumder, and Sri. Abdur Rehman, all of whom have helped in various ways too numerous to mention.

I also extend my appreciation to the members of Artech House for their keen interest in this project. Finally, I am grateful to my sons, Dr. Souvik Sarkar and Sri. Suryaday Sarkar, without whose understanding such a major project and its successful completion would not have been happened. I am not expressing my thankfulness to my wife, Smt. Bani Sarkar, because according to Hindu tradition husband and wife are a single entity.

Finally, I shall feel amply rewarded of my labor if the present book proves useful to those for whom it is meant. No doubt, the road to perfection leads to infinity; still there is always room for improvement. Any constructive criticism and suggestions for the improvement of the book will be acknowledged.

1
Introduction

1.1 Overview of Wireless Networks

Recent advancements in information technology and continued miniaturization of mobile communication devices have increased the use of wireless communication environments manifold. Because of this tremendous growth, wireless networks have witnessed rapid changes and the development of new applications. Recently, wireless networks have come into prominence, and the impact of wireless networks has been and will continue to be profound because they hold the potential to revolutionize many segments of our economy and life, from environmental monitoring and conservation to manufacturing and business asset management and to automation in the transportation and health care industries. Practically, very few inventions have been able to "shrink" the world in such a manner. The standards that define how wireless communication devices interact are quickly converging and allow the creation of a global wireless network that delivers a wide variety of services. Wireless networks are characterized by dynamic topologies, bandwidth-constrained variable-capacity wireless links, energy-constrained operations, and limited physical security. Wireless networks are particularly useful for providing communication support where no fixed infrastructure exists or where the deployment of a fixed infrastructure is economically not feasible.

Wireless networks provide users with seamless access to information, like checking email, browsing the Internet, and accessing corporate data without getting disconnected from the network even when the user is in motion. Wireless technologies have enabled us to lead a convenient way of life by becoming part and parcel of our daily routine. Users can obtain information in real time from various sources to meet their needs. Wireless solutions give a competitive edge for a user over his rivals. In major cities we can find "hot spots" where Internet services are provided to users with Internet-enabled wireless devices. Commonly used devices for wireless networking include tablet PCs, personal digital assistants (PDAs), cellular phones, various sensors, and laptops [1–10].

Wireless networks are easy to set up, as there is no need for cable installations, thereby increasing the reliability of the network. Wireless networks allow us to extend our accessibility to places where wiring is nearly impossible. Wireless networks can be put into operation by various devices in a typical office or campus setup. A typical wireless setup in an office environment is shown in Figure 1.1.

The differences between wired and wireless networks are as follows:

- In wired networks the devices are attached to cables, which make us immobile, while wireless networks offer us the convenience of mobility.
- Wired networks can be costly for cables and other installation equipment when covering wide geographical areas, whereas wireless networks do not incur this cost.
- Wired networks have higher transmission speed than that of wireless networks.

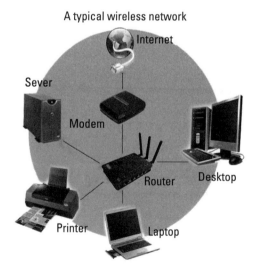

Figure 1.1 A typical wireless network setup. (© http://www.osd.noaa.gov/TPIO/Freq-Mang/freq_mang.html)

- In a wired network, a user gets high speed with a dedicated line, but in wireless networks the connection speed may be less as the network may be used by multiple users.

1.2 Types of Wireless Communication

Wireless communication can be classified into wireless personal area networks (WPANs), wireless local area networks (WLANs), wireless metropolitan area networks (WMANs), and wireless wide area networks (WWANs) based on the communication range (Figure 1.2).

1.2.1 Wireless Personal Area Networks Bluetooth

In WPAN, as the name itself indicates, various devices are connected around an individual's personal area. Connection can be established in a peer-to-peer ad hoc fashion. An individual's personal space is the maximum distance (~10 meters) up to which a communication is possible with the concerned wireless devices. Wireless sensor networks, Bluetooth, and infrared are the prominent WPAN technologies. IEEE has established the IEEE 802.15 Working Group to ratify the standards for WPAN technologies [11–16].

1.2.2 Wireless Local Area Networks 802.11

In WLAN the connection is established within a local area. Here the local area can be campus buildings, corporate buildings, an airport, or a railway station [6]. WLAN is more useful in places where extensive wiring is prohibited, like in museums, archaeological buildings, or in offices. WLAN can operate in two modes, namely infrastructure connected and purely infrastructureless. In infrastructure WLAN the wireless devices connect to an access point, which in turn is connected to an existing Internet backbone. In purely infrastructure mode the devices are connected in a peer-to-peer ad hoc fashion. Here connection is established temporarily between the devices without using any access point. IEEE has approved IEEE 802.11 as the standard to be used with WLANs. Wi-Fi and mobile ad hoc networks are the prominent technologies under WLANs [17–25].

1.2.3 Wireless Metropolitan Area Networks WiMAX

WMAN is also known as wireless in local loop (WLL). WMANs enable users to connect to multiple networks in different buildings, stadiums, corporate houses, or hotels within a metropolitan area. WMAN provides an alternate to leased lines and fiber or copper cabling. IEEE 802.16 is the work group as-

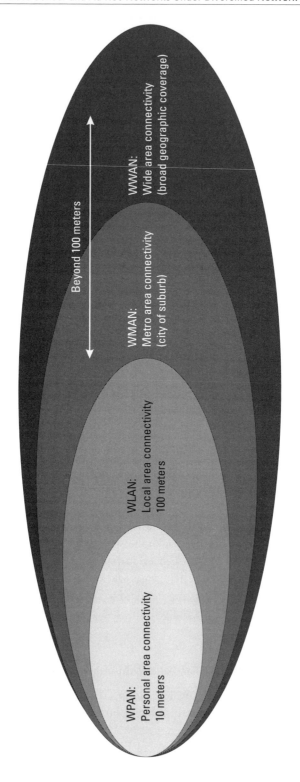

Figure 1.2 Categories in wireless communication.

signed by IEEE to standardize the specifications to enable wireless broadband using WMANs. WiMAX is the prominent technology under WMAN [26–31].

1.2.4 Wireless Wide Area Networks

Wireless wide area networks are hardly new. These networks have been utilized since the mid 1980s, when microwave transmissions were beamed about by complex and powerful transmitting units that needed Federal Communications Commission licenses and radios and antennas. Presently, wireless systems can deliver up to 100 Mbps speeds at 40 miles distance, and speeds are increasing. A wide area network (WAN) is a communication network that utilizes devices such as telephone lines, satellite dishes, antennas, and microwaves to span a larger geographic area than can be covered by a local area network (LAN).

WWANs are established over large geographical areas like cities or countries. Connections are established through the use of mobile phone signals provided by satellites or multiple high beaming antennas. Prominent technologies under WWAN are global system for mobile communication (GSM), general packet radio service (GPRS), and code division multiple access (CDMA).

Since radio communication systems do not provide a physically secure connection path, wireless WANs typically incorporate encryption and authentication methods to make them more secure. Unfortunately some of early GSM encryption techniques were flawed, and security experts have raised warnings that cellular communication, including WWAN, is no longer secure.

The WWAN can access the Internet through cellular technologies. These networks provides fast data speeds compared to mobile technologies, and their range is also extensive. Their network connectivity allows a user with laptop and WWAN card to browse the Internet , check emails, or connect to a virtual private network (VPN) from anywhere within the geographical range of cellar service. As more and more people depend on Internet technology to conduct business and keeps information flowing, wireless connectivity has become a virtual necessity. WWAN guarantees connectivity when it is necessary, and it is available in regions where services like digital subscriber line (DSL) and cable might not be [32–36].

Apart from these wireless networks, there are two more special types of wireless networks—ad hoc and wireless sensor networks—discussed extensively in this book.

1.3 The IEEE 802.11 Family

The IEEE 802.11 is a standard set and maintained by the IEEE LAN/MAN Standards Committee (IEEE 802) and Part 11: Wireless LAN Medium Access

Control (MAC) and Physical Layer (PHY) Specifications. This standard is considered for implementing information exchange between systems—a wireless client and a base station or between two wireless clients. The IEEE accepted the specification in 1997. IEEE 802.11 is one of the earliest technologies to be adapted as the standard for manufacturing wireless devices by various vendors and has iterated in various forms since its inception.

Some of the several specifications in the 802.11 families are as follows:

- 802.11: This standard applies to wireless LANs and provides 1 or 2 Mbps transmission in the 2.4-GHz band using either frequency hopping spread spectrum (FHSS) or direct sequence spread spectrum (DSSS).
- 802.11a: This standard is an extension to 802.11 that applies to wireless LANs and provides up to 54 Mbps in the 5-GHz band. 802.11a uses an orthogonal frequency division multiplexing (OFDM) encoding scheme rather than FHSS or DSSS.
- 802.11b: This standard, also referred to as 802.11 high rate or Wi-Fi, is an extension to 802.11 that applies to wireless LANS and provides 11 Mbps transmission (with a fallback to 5.5, 2, and 1 Mbps) in the 2.4-GHz band. 802.11b uses only DSSS. 802.11b was ratified to the original 802.11 standard in 1999, allowing wireless functionality comparable to Ethernet.
- 802.11g: This standard applies to wireless LANs and is used for transmission over short distances at up to 54 Mbps in the 2.4-GHz bands.
- 802.11n: This standard builds upon previous 802.11 standards by adding multiple-input multiple-output (MIMO) for increased data throughput through spatial multiplexing and increased range by exploiting the spatial diversity through several coding schemes. The real speed would be 100 Mbps or even 250 Mbps in PHY level and so up to four to five times faster than 802.11g.

The 802.11 family consists of several modulation techniques that use the same basic protocol. The most popular of those defined here are 802.11b and 802.11g protocols, which are amendments to the original standard. 802.11-1997 was the first wireless networking standard, but 802.11b was the first widely accepted one, followed by 802.11g and 802.11n. 802.11n is a new multistreaming modulation technique.

The differences between different flavors of IEEE 802.11 standard discussed [37–41] are shown in Table 1.1.

Table 1.1
Various IEEE 802.11 Standards

Standard	802.11 (Base Standard)	802.11a	802.11b	802.11g	802.11n
Standardized year	1997	1999	1999	2003	2009
Frequency band	2.4 GHz	5.8 GHz	2.4 GHz	2.4 GHz	2.4–5.8 GHz
Maximum range	~70 m	~100 m	~100 m	~110 m	~160 m
Maximum bandwidth	2 Mbps	54 Mbps	11 Mbps	54 Mbps	248 Mbps
Spread spectrum technology	DSSS, FHSS	OFDM	DSSS, CCK	OFDM	MIMO

1.4 Cellular Network

The cell phone network is the most common example of a cellular network. A cell phone is a portable phone that receives or makes calls through a cell site or transmitting tower. Radio waves are employed to transfer signals to and from the cell phone. It is a radio network distributed over land areas called cells, each served by at least one fixed-location transceiver known as a base station (BS). These cells provide radio coverage over a wide geographical area when joined together. This supports a large number of portable transceivers to communicate with each other and with fixed transceivers and telephones anywhere in the network, through cell sites, even if some of the transceivers are moving through more than one cell during transmission. The mobile phone operators use cellular networks to achieve both coverage and capacity for their subscribers. In order to avoid line-of-sight signal loss and to provide service support to a large number of active phones, large geographic areas are divided into smaller cells in that area. All of the cell sites are linked to telephone exchanges, and these exchanges connect to the public telephone network. Figure 1.3 shows a cellular network.

The distinct advantages of cellular networks are listed as follows:

- Reduced interference from other signals;
- Increased capacity;
- Larger coverage area;
- Reduced power use.

1.4.1 Existence of Mobile Generation and 4G

In the last two decades, a phenomenal growth in wireless communication, especially mobile communications systems, has revolutionized the way people communicate. Looking back to history, wireless access technologies have followed different evolutionary paths aiming for performance and efficiency in

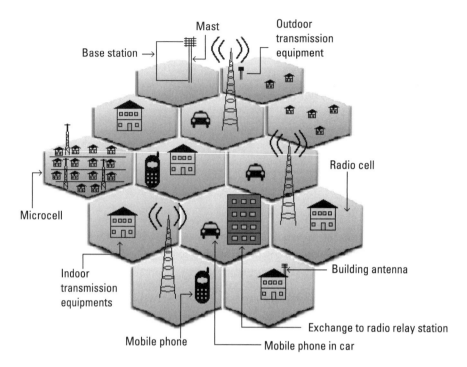

Figure 1.3 Cellular network. (© http://bulbing.com/projects/cellular.html.)

high mobile environment. This section provides an overview of the evolution of mobile wireless communication networks.

There has been a clear shift from fixed to mobile cellular telephony, especially since the turn of the year 2000. Many designing scenarios have developed with not only 2G networks but also the evolution of 2G to 2.5G or even to 3G networks. Along with this, interoperability of the networks have to be considered. 1G refers to analog cellular technologies with voice mobility; it became available in the 1980s. 2G denotes initial digital systems, introducing services such as short messaging and lower speed data. CDMA2000 and GSM are the primary 2G technologies, although CDMA2000 is sometimes called a 3G technology because it meets the 144-Kbps mobile throughput requirement. EDGE also meets this requirement. 2G technologies became available in the 1990s. 3G requirements were specified by the ITU as part of the International Mobile Telephone 2000 (IMT-2000) project, for which digital networks had to provide 144 Kbps of throughput at mobile speeds, 384 Kbps at pedestrian speeds, and 2 Mbps in indoor environments. WiMAX is designated as an official 3G technology. The deployment of WiMax started in the last decade. The ITU has recently issued requirements for IMT-Advanced, which constitutes the official definition of 4G. Requirements include operation in up to 40-MHz radio channels and extremely high spectral efficiency. The ITU recommends operation in up to 100-MHz radio channels and peak spectral efficiency of 15 bps/Hz, resulting in a theoretical throughput rate of 1.5 Gbps. Previous to the publication of the requirements, 1 Gbps was frequently cited as a 4G goal. No available technology meets these requirements, and new technologies like LTE-Advanced and IEEE 802.16m will be required. Some have tried to label current versions of worldwide interoperability for microwave access (WiMAX) and long term evolution (LTE) as 4G, but this is only accurate to the extent that such designation refers to the general approach or platform that will be enhanced to meet the 4G requirements. With WiMAX and high speed packet access (HSPA) significantly outperforming 3G requirements, calling these technologies 3G clearly does not give them full credit, as they are a generation beyond current technologies in capability. But calling them 4G is also not correct. Unfortunately, the generational labels do not properly capture the scope of available technologies and have resulted in some amount of confusion. In contrast to 3G, the new 4G framework to be established will try to accomplish new levels of user experience and multiservice capacity by integrating all the mobile technologies that exist. The reason for the transition to the all-IP is to have a common platform for all the technologies that have been developed so far. This means that this network will be less expensive and data transfer will be much faster. The 4G mobile communication services started in 2010 and are expected to become mass market in about 2014–2015; then the user has the freedom and flexibility to select any desired service with reasonable QoS and affordable price anytime and anywhere.

1.4.2 WiMAX

WiMAX is a telecommunication protocol created and promoted by WiMAX Forum for Conformity and Interoperability of the standard that provides fixed and mobile Internet access. The current WiMAX revision provides up to 40 Mbps with the IEEE 802.16m update expected to offer up to 1 Gbps fixed speeds. The forum describes as a standard-based technology alternative to cable and DSL.WiMAX refers to interoperable implementations of the IEEE 802.16 wireless-networks standard, which is IEEE 802.11 wireless LAN standard.

The IEEE 802.16 standard forms the basis of WiMAX, and it is sometimes referred as WiMAX, Fixed WiMAX, Mobile WiMAX, 802.16d, and 802.16e. The IEEE 802.16d is sometimes referred to as Fixed WiMAX, since it has no support for mobility. The IEEE 802.16e introduces support for mobility, among other things, and is therefore also known as Mobile WiMAX.

WiMAX is designed to support high-speed wireless broadband Internet access on the move and can be seen as the improved version of Wi-Fi. It is intended to coexist with Wi-Fi. In WiMAX, line of sight is not required between a subscriber device and the base station that operates at an expected throughput of 15 Mbps. The line-of-sight communication takes place between WiMAX antennas at an expected throughput of 40 Mbps. Thousands of users can be supported from a single base station. WiMAX offers higher bandwidth and range than Wi-Fi and provides a bandwidth of 70 Mbps. WiMAX can cover a distance of more than 50 km due to an increase in the power of the transmitter and can operate in low frequency band of 2–2.5 GHz. The coding technique used is the demand assignment multiple access–time division multiple access (DAMA-TDMA). WiMAX provides an alternative to leased lines and is highly dependable in terms of security. By having a cluster of WiMAX transmitters, information can be accessed even from far off places. Quality of service is a main concern in WiMAX technology. People tend to fear that an increase in the number of users using the same tower might result in congestion. But there are many built-in mechanisms to ensure that if extra users try to access the same saturated tower, then those extra users are automatically transferred to another WiMAX tower. Figure 1.4 shows the typical WiMAX.

The bandwidth and range of WiMAX make it suitable for many potential applications, such as:

- Providing portable mobile broadband connectivity across cities and countries;

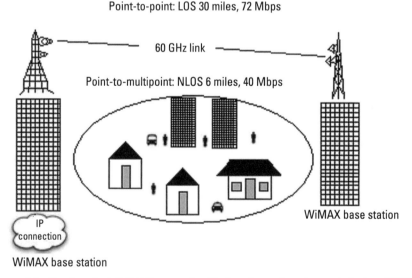

Figure 1.4 Fixed typical WiMAX scenario. (© http://www.laercio.com.br/artigos/HARDWARE/hard-107/hard-107.htm.)

- Providing a wireless alternative to cable and DSL operator;
- Providing data, telecommunications (voice over Internet protocol, VoIP), and Internet protocol television (IPTV) services.

WiMAX is a possible replacement candidate for cellular mobile technologies such as GSM and CDMA, and it supports such services as quality of service (QoS) and multicasting [42–45].

1.5 Mobile Ad Hoc Networks

A mobile ad hoc network is an autonomous network of mobile nodes with no fixed or centralized infrastructure (Figure 1.5). These mobile nodes can join or leave the network with changes in topology. If the destination node is out of reach from the source node, then the packets have to be routed through various intermediate nodes by applying an appropriate routing algorithm. Each of the nodes acts as a host as well as a router for forwarding the packets. Ad hoc networks have lower bandwidth compared to wired networks. Even though an ad hoc network works in isolation from other networks, it can connect to Internet through gateways. Ad hoc networks can be applied in search and rescue operations, classrooms, video conferencing, and disaster areas [46–54]. A mobile ad hoc network (MANET) consists of a group of wireless mobile nodes that dynamically self organize and self control to communicate with each other without infrastructure. Each node acts as a transceiver, and communicates via hops. When some nodes are out of the coverage area due to high mobility, the topology of the networks changes dynamically, thus making routing very challenging. The mobile ad hoc network is dealt in great detail in Chapter 2.

1.6 Wireless Sensor Networks

In recent years, advancement in semiconductor technology and miniaturization, simple low-power circuit design, and improved cheaper and smaller batteries made possible wireless sensor networks (WSNs). These networks combine wireless communication and minimum computational facilities with sensing of physical phenomena, which can easily embed in our physical environment. It is expected that the size of a sensor will be a few cubic millimeters and the price less than one U.S. dollar, including radio front end, microcontroller, power supply, and the actual sensor. All these components embedded together in a single device form a *sensor node*. In other words, a sensor node is basically a device that converts sensed attributes (such as temperature, pressure, and vibrations) into a form understandable by the users.

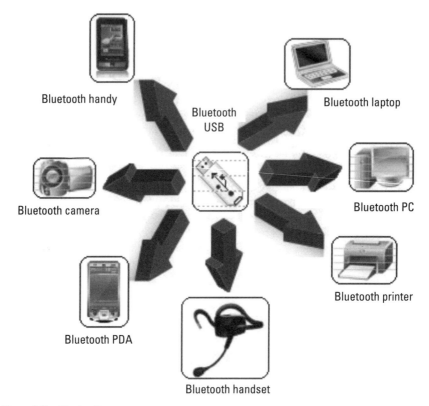

Figure 1.5 Bluetooth.

The popularity of sensor networks can be gauged at the number of research papers being published in premier conferences sponsored by various organizations like the Institute of Electrical and Electronics Engineers (IEEE) and the Association for Computing Machinery (ACM). WSNs, which can be considered special case ad hoc networks with reduced or no mobility, are expected to find increasing deployment in the coming years, as they enable reliable monitoring and analysis of unknown and untested environments. These networks are data centric (i.e., unlike traditional ad hoc networks, where data is requested from a specific node, the requested data is based on certain attributes). Therefore a large number of sensors need to be deployed to accurately reflect the physical attribute of a given area. There are few integrated sensors commercially available that can be used directly as plug-and-play units to monitor and control all the physical parameters required by the user. But there are many basic sensors transducers that could convert many physical quantities, such as temperature, pressure, velocity, acceleration, stress and strain, light, sound, light intensity, humidity, and gas sensors. These basic generic transducers need to be interfaced and connected to other devices, and such custom units can be used for a given specific application.

Sensor networks consist of a large number of sensor nodes that can be scattered over the phenomenon requiring study. Each of these sensor nodes have limited battery power and have to perform multihop communication to send the data to the sink node or base station. With the improvement of technology, sensor nodes can be produced in huge quantities at a very low cost; thereby the applicability of sensor networks tends to get more feasible. Sensor nodes are deployed in huge quantities and are generally dense, resulting in a high degree of tolerance. Failure of one or two nodes among a million nodes does not affect the overall data collection mechanism of sensor networks. A sensor network is fully void of human interference and is unattended once it is deployed. The routing problem in sensor network is different from that of an ad hoc network due to the increase in density of sensor nodes and because sensor nodes are very much resource constrained, since they are battery operated. Sensor networks can be used in military operations for tracking enemy vehicle movement as well as enemy soldier movement, in monitoring nuclear power installations, in controlling the climatic ability in buildings, in monitoring the soil conditions for agricultural needs, in monitoring the seismic activities in sensitive areas, and in monitoring the health of patients [55–62]. The various research issues, challenges, difference between ad hoc networks and sensor networks, protocols, hardware, deployment, and applications of sensor networks are discussed at length in Chapter 6.

1.7 Bluetooth

Bluetooth technology was developed at Ericsson. Bluetooth can be used to form wireless personal area networks having a short range of communication. Some of the features of Bluetooth technology are that it is robust, consumes less power, and costs less to manufacture. Bluetooth can be found in a number of devices, including computer peripherals, digital cameras, and GPS devices. Bluetooth can be configured as a piconet or scatternet. In a piconet, eight devices communicate with each other using FHSS. Among them, one device acts as master device and the remaining seven devices connect to this master device. Collisions do not occur in a piconet as the master device assigns a communication slot for each of the seven slave devices. A scatternet is made up of clusters of piconets. A master device in one piconet can act as a slave in another piconet. The master device plays the role of a bridge in connecting two piconets. Bluetooth operates at 2.4-GHz frequency range. The maximum data rate achievable by Bluetooth devices is up to 1 Mbps. Some of the applications of Bluetooth technology are as follows: (a) it allows the transfer of voice and images between devices that are portable and stationary, (b) it eliminates the need to connect the devices using cable, as devices can communicate with each other up a range of 10m, and (c) connection is established between Bluetooth devices as soon as they come within their transmission range. Some of the disadvantages of Bluetooth connection are as follows: (a) it has slow transfer rate, and (b) it is susceptible to interference.

Bluetooth, a standard wire-replacement communications protocol, is basically designed for low-power consumption with a short range based on low-cost transceiver microchips in each device. With change of classes, maximum permitted power and range vary from 100 mW and 100 meters to 1mW and 1meter. The range is power class dependent, but the effective ranges vary in practice due to propagation conditions, material coverage, production sample variations, antenna configurations, and battery conditions. The Bluetooth has several versions with varying data rates. Version 1.2 has data rate of 1 Mbps with maximum application throughput of 0.7 Mbps, whereas version 2.0+EDR has a 3-Mbps data rate with maximum application throughput of 2.1 Mbps. Other versions have also become available with high data rates, like 24 Mbps [63–71]. Figure 1.5 shows the typical Bluetooth setup.

1.8 Organization of the Book

The following chapters of this book are organized in the following manner:

- In Chapter 2, a general introduction and overview of ad hoc networks is presented. In this chapter we have identified the core areas of ad hoc networks. We give a preview of each of these areas.
- In Chapter 3, an exhaustive description of various routing protocols in ad hoc networks is given. Classification of routing protocols and various characteristics in designing routing protocols have also been touched upon. We have studied 11 routing protocols that have been used at various stages in this book. We have compared six different routing protocols, namely ad hoc on-demand multipath distance vector (AOMDV), dynamic MANET on demand (DYMO), fisheye state routing (FSR), location-aided routing protocol (LAR), optimized link state routing protocol (OLSR), and temporarily ordered algorithm (TORA), using community based mobility model and SMS mobility model. Of these, the LAR, AOMDV, and TORA routing protocols are analyzed with the community mobility model, and OLSR, DYMO, and FSR protocols are analyzed using the SMS mobility model. Also, we have compared AOMDV and OLSR routing protocols using the Levy walk mobility

model and Gauss-Markov mobility model. A thorough investigation of these routing protocols over mobility models, which reflect real-world scenarios, are very important as the routing protocols affect the overall performance of the network.

- In Chapter 4, we have elaborated on the experimental evaluation of the mobility model in ad hoc networks. Mobility models play a very important role in simulating the routing protocols. Eleven mobility models have been reviewed and classified according to their nature. With the help of various metrics, these mobility models have been simulated and evaluated. In this chapter we have carried out performance comparison of various mobility models like community model, group force mobility model (GFMM), reference point group mobility (RPGM), Manhattan mobility model and random waypoint-steady state (RWP-SS) mobility model, random waypoint (RWP), random walk with reflection (RW-R), and random walk with wrapping (RW-W). The community model, GFMM, and RPGM mobility models are pure group mobility models, while the Manhattan mobility model can be considered a pseudo group mobility model. Also we have compared mobility models like random waypoint, random walk with reflection, and random walk with wrapping.

- In Chapter 5, overhead related issues are presented. In particular, energy overhead along with energy consumption patterns of various routing protocols under different energy models is studied. We have studied the energy overhead performance of three different routing protocols under three different energy models. The three different energy models considered are (a) Bansal energy model, (b) Vaddina energy model, and (c) Chandrakasan energy model. We apply these energy models to the AOMDV, TORA, and OLSR routing protocols to determine the energy overhead among these three routing protocols by varying the transmission range. We have analyzed the energy consumption of five mobility models under five different routing protocols. The mobility models considered are RW-W, RWP-SS, Gauss-Markov mobility model (GM), community based mobility model (CM), and semi-Markov smooth mobility model (SMS). The selected routing protocols are AOMDV, DYMO, FSR, OLSR, and TORA.

- In Chapter 6, a comprehensive study of wireless sensor networks is done. Wireless sensor networks have become a hot research theme in academia and as well as in industry in recent years due to its wide range of applications, from medical research to military. Sensor networks have been stimulated by the need for setting up the communication networks to gather information in situations where a fixed infrastructure cannot be employed on the fly, as it occurs in the management of emergencies and disaster recovery. Various challenging issues, applications, and research directions are identified and discussed in detail in this chapter. Also, a wireless sensor and actor network scenario are evaluated using different routing protocols.

- In Chapter 7, a comprehensive analysis of the effect of IEEE 802.15.4 MAC protocol on the performance of ZigBee routing protocol is evaluated with static as well as mobile nodes in wireless sensor networks. The IEEE 802.15.4 is the standard adopted for the wireless sensor network platform. AODV is defined as the underlying routing protocol in ZigBee. Different topologies, low-rate wireless personal area networks (LR-WPAN) architecture, superframe structure, data transmission, and slotted carrier sense multiple access collision (CSMA/CA) mechanisms of IEEE 802.15.4 are discussed in detail.

- In Chapter 8, various patrimonial ZigBee routing protocols are evaluated, and the effect of traffic on the performance of a wireless sensor network under various traffic generators is studied in detail. In this work we study the exponential on/off traffic effect on the performance of wireless sensor networks using the IEEE 802.15.4/ZigBee standard. There is a need to understand the effect of traffic on versatile behavioral aspects of wireless sensor networks, like number of sources and sensor nodes. This study is the first of its kind to compare the performance of the AODV family of routing protocols, namely AODV, AODVUU, AOMDV, and RAODV routing protocols, from a sensor network point of view.

- In Chapter 9, the various applications and recent developments is discussed elaborately with opportunities and future trends, along with some typical applications in the area of ad hoc network and sensor networks including recent developments, research trends, challenges, and future scope.

References

[1] Willig, A., K. Matheus, and A. Wlisz, "Wireless Technology in Industrial Networks." In *Proceedings of the IEEE*, Vol. 93, No. 6, June 2005, pp. 1130–1151.

[2] Benker, Y., "Some Economies of Wireless Communications," *Harvard Journal of Law and Technology*, Vol. 16, No. 1, Fall 2002.

[3] Zhang, Q., and Y.-Q. Zhang, "Cross Layer Design for QoS Support in Multihop Wireless Networks," *Proceedings of the IEEE Communications*, Vol. 96, No. 1, 2008.

[4] Lee, M.-H., and H. Yoe, "Comparative Analysis and Design of Wired and Wireless Integrated Networks for Wireless Sensor Networks." In *Proceedings of the 5th ACIS International Conference on Software Engineering Research, Management, and Applications*, Busan, August 20–22, 2007.

[5] Li, P., and Y. Fang,"The Capacity of Heterogeneous Wireless Networks." In *Proceedings of the IEEE INFOCOM*, March 14–19, 2010, San Diego, CA.

[6] Zhao, W., M. Ammar, and E. Zegura, "The Energy Limited Capacity of Wireless Networks." In *Proceedings of the First Annual IEEE Communications Society Conference on Sensor and Ad Hoc Communications and Networks*, October 4–7, 2004.

[7] Zheng, Z., et al., "A Study of Network Throughput Gain in Optical Wireless (FiWi) Networks Subject to Peer-to-Peer Communications." In *Proceedings of IEEE International Conference on Communications (ICC'09)*, Dresden, Germany, June 14–18, 2009.

[8] Aspas, J. P., and C. G. Montenegro, "Focusing the Study on Wireless Multihop Networks." In *Proceedings of the 13the International Telecommunications Network Strategy and Planning Symposium*, Budapset, September 28–October 2, 2008.

[9] Ma, L., and J. Dongyan, "The Competition and Cooperation of WiMAX, WLAN and 3G," 2nd International Conference on Mobile Technology, Applications, and Systems, November 15–17, 2005, pp. 1–5.

[10] Heftman, G., "Wireless Up and Comers Head for the Spotlight," *Microwaves and RF*, April 2000.

[11] Karaoguz, J., "High-Rate Wireless Personal Area Networks," *IEEE Communications Magazine*, February 2004, pp. 132–138.

[12] Neelakanta, P. S., "Designing Robust Wireless Communications for Factory Floors." In *Proceedings of 2006 IEEE International Conference on Industrial Informatics*, Singapore, August 16–18, 2006.

[13] Ghetie, J., "Fixed Wireless and Cellular Mobile Convergence: Technologies, Solutions and Services." In *Proceedings of 9th International Conference on Telecommunications 2007*, Zagreb, Croatia, June 13–15, 2007.

[14] Kim, S., I. Nam, et al., "A Single Chip 2.4-GHz Low Power CMOS Receiver and Transmitter for WPAN Applications." In *Proceedings of Radio and Wireless Conference (RAWCON'03)*, August 10–13, 2003.

[15] Jeon, W.-G., Y. H. You, et al., "Timing Synchronization for IEEE 802.15.3 WPAN Applications," *IEEE Communications Letters*, Vol. 9, No. 3, March 2005.

[16] Hasbollah, A. A., S. H. S Ariffin, et al., "Performance Analysis for 6loWPAN IEEE 802.15.4 with IPv6 Network," 2009 IEEE Region 10 Conference (TENCON2009), Singapore, January 23–26 2009.

[17] U.S. Robotics, "Wireless LAN Networking," White Paper, pp. 1–11.

[18] Sarkar, N. I., "WLAN Designer: A Web-Based Software Tool To Enhance Teaching and learning wireless LAN Design." In *Proceedings of IEEE International Conference on Advanced Learning Technologies*, August 30–September 1, 2004.

[19] Janevski, T., A. Tudzarov, et al., "Applicative Solution for Easy Introduction of WLAN as Value Added Service in Mobile Networks." In *Proceedings of Vehicular Technology Conference 2007*, Dublin, Ireland, April 22–25, 2007.

[20] Senkindu, S., and H. A. Chan, "Enabling End to End Quality of Service in a WLAN Wired Network." In P*roceedings of IEE Military Communications Conference (MILCOM 2008)*, San Diego, CA, November 16–19, 2008.

[21] Guo, F., and T. Chiueh, "Scalable and Robust WLAN Connectivity Using Access Point Array." In *Proceedings of International Conference on Dependable Systems and Networks, 2005*, June 28–July 1, 2005.

[22] Jo, J. H., and H. Jayant, "Performance Evaluation of Multiple IEEE 802.11b WLAN Stations in the Presence of Bluetooth Radio Interference." In *Proceedings of the IEEE International Conference on Communications (ICC'03)*, May 11–15, 2003.

[23] Huang, W. H., K.-C. Tang, et al., "The Experimental Campus WLAN Roaming System and WiMax Inegration in Taiwan." In *Proceedings of 14th Asia-Pacific Conference on Communications 2008*, Tokyo, October 14–16, 2008.

[24] Soni, V., and R. Mendratta, "Next Generation WLAN Architecture for High Performance Networks," IET International Conference on Wireless, Mobile, and Multimedia Networks," Mumbai, India, January 11–12, 2008.

[25] Sadhukhan, S. K., S. Mandal, and D. Saha, "A Practical Approach for Planning WLAN Deployment Under 3G Cellular Network." In *Proceedcings of First International Conference on Networks and Communications*, Chennai, India, December 27–29, 2009.

[26] Yahiya, A. T., K. Sethom et al., "A Case Study: IEEE 802.21 Framework Design for Service Continuity across WLAN and WMAN." In *Proceedings of International Conference on Wireless and Optical Communications Networks*, Singapore, July 2–4, 2007.

[27] Yahiya, A., and G. Pujolle, "Seamless Continuity of Service Across WLAN and WMAN Networks: Challenges and Performance Evaluation." In *Proceedings of the 2nd International IEEE/IFIP International Workshop on Broadband Convergence Network*, Munich, May 21–21, 2007.

[28] Agrawal, D. P., and H. Gossain, "Recent Advances and Evolution of WLAN and WMAN Standards," *IEEE Wireless Communications*, Vol. 15, No. 5, October 2008.

[29] Reinwand, C. C., "Municipal Broadband: The Evolution of Next Generation Wireless Networks." In *Proceedings of IEEE Radio and Wireless Symposium*, Long Beach, CA, January 9–11, 2007.

[30] Ghazisaidi, N., and H. Kassaei, "Integration of WiFi and WiMax-Mesh Networks." In *Proceedings of 2nd International Conference on Advances in Mesh Networks*, Athens and Glyfada, Greece, June 18–23, 2009.

[31] Nie, J., and H. Xin, "Communication with Bandwidth Optimization in IEEE 802.16 and IEEE 802.11 Hybrid Networks." In *Proceedings of IEEE International Symposim on Communications and Information Technology*, October 12–14, 2005.

[32] Pack, S., and H. Rutagewma, "An Integrated WWaN-WLAN Link Model in Mobile Hotspots." In *Proceedings of Internatioanl Conference on Communications and Networking in China*, Beijing, China, October 25–27, 2006.

[33] Pack, S., H. Rutagemwa, X. Shen, J. W. Mark, et al., "Performance Analysis of Mobile Hotspots with Heterogeneous Wireless Links," *IEEE Transactions on Wireless Communications*, Vol. 6, No. 10, October 2007.

[34] Suh, S. Y., and S. L. Ooi, "Challenges on Multi Radio Antenna Systems for Mobile Devices." In *Proceedings of 2007 IEEE International Symposium on Antennas and Propagation Society*, Honolulu, HI, June 9–15, 2007.

[35] Wei, H. Y., and R. D. Gitlin, "WWAN/WLAN Two Hop Relay Architecture for Capacity Enhancement." In *Proceedings of 2004 IEEE Wireless Communication and Networking Conference*, March 21–25, 2004.

[36] Marihart, D. J., "Overview of Mobile Computing Technology including 3G and 4G." In *Proceedings of IEEE Power Engineering Society Summer Meeting*, Vancouver, BC, 2001.

[37] Carrano, R. C., L. C. S. Magalhães, D. C. Muchaluat-Saade, C. V. N. Alburquerque, et al., "IEEE 802.11s Multihop MAC: A Tutorial," *IEEE Communication Surveys and Tutorials*, Vol. 13, No. 1, First Quarter 2011, pp. 52–67.

[38] Jiang, R., and V. Gupta, "Interaction Between TCP and the IEEE 802.11 MAC protocol." In *Proceedings of the International Conference of DARPA Information Survivability Conference and Exposition*, April 22–24, 2003.

[39] Zhu, H., and G. Cao, "On Improving the Performance of IEEE 802.11 with Multihop Concepts." In *Proceedings of the 12th International Conference on Computer Communications and Networks*, October 20–22, 2003.

[40] Wen Kuang and Kuo C., "Enhanced Backoff Scheme in CSMA/CA for IEEE 802.11." In *Proceedings of International Vehicular Technology Conference*, October 6–9, 2003.

[41] Brenner, P., "A Technical Tutorial on the IEEE 802.11 Protocol," Breezecom Wireless Communications, July 1996, pp. 1–24.

[42] Abichar, Z., Y. Peng, and J. M. Chang, "WiMAX: The Emergence of Wireless Broadband," IEEE Computer Society, IT Professional, July–August 2006, Vol 8 (4), pp. 44–48.

[43] Vaughan-Nichols, S. J., "Achieving Wireless Broadband With WiMAX," *IEEE Industry Trends*, June 2004, Vol. 37 (6), pp. 10–13.

[44] Ghosh, A., and D. R. Wolter, "Broadband Wireless Access with WiMAX/802.16: Current Performance Benchmarks and Future Potential," *IEEE Communications Magazine*, February 2005, Vol. 43, No. 2, pp. 129–136.

[45] Bing, B., "Broadband Wireless Access-The Next Wireless Revolution," *Proceedings of the 4th Annual Communication Networks and Services Research Conference (CNSR'06)*, Moncton, Canada, March 24–25, 2006.

[46] Chadha, R., "Managing Mobile Ad hoc Networks." In *Proceedings of IEEE/IFIP Network Operations and Management Symposium*, Seoul, South Korea, April 23, 2004.

[47] Maihofer, C., "A Survey of Geocast Routing Protocols," *IEEE Communications Surveys and Tutorials*, Vol. 6, No. 2, 2004, pp. 32–42.

[48] Zhang, Z., "Routing in Intermittently Connected Mobile Ad Hoc Networks and Delay Tolerant Networks: Overicew and Challenges," *IEEE Communications Surveys and Tutorials*, Vol. 8, No. 1., 2006, pp. 24–37.

[49] Hanbali, A., and E. Altman, "A Survey of TCP over Ad Hoc Networks," *IEEE Communications Surveys and Tutorials*, Vol. 7, No. 3, 2005, pp. 22–36.

[50] Abduljalil, F. M., and S. K. Bodhe, "A Survey of Integrating IP Mobility Protocols and Mobile Ad Hoc Networks," *IEEE Communications Surveys and Tutorials*, Vol. 9, No. 1, 2005, pp. 22–36.

[51] Lima, M., A. des Santos, and G. Pujolle, "A Survey of Survivability in Mobile Ad Hoc Networks," *IEEE Communications Surveys and Tutorials*, Vol. 11, No 1, 2009, pp. 66–67.

[52] Junhai, L., Y. Danxia, et al., "A Survey of Multicast Routing Protocols for Mobile Ad Hoc Networks," *IEEE Communications Surveys and Tutorials*, Vol. 11, No. 1, 2009, pp. 78–91.

[53] Rrodigh, M., et al., "Wireless Ad Hoc Networking: The Art of Networking Without a Network," *Ericsson Review*, No. 4, 2000.

[54] Lin, T., et al., "A Framework for Mobile Ad hoc Routing Protocols," In *Proceedings of IEEE 2003 Wireless Communication and Networking Conference (WCNC 2003)*, 2003.

[55] Fowler, K. R., "The Future of Sensors and Sensor Networks Survey Results Projecting the Next Five Years." In *Proceedings of IEEE Sensor Application Symposium, 2009*, New Orleans, LA, February 17–19, 2009.

[56] Ahmad, A. A., H. Shi, and Y. Shang, "A Survey on Network Protocols for Wireless Sensor Networks." In *Proceedings of International Conference on Information Technology, Research, Education*, August 11–13, 2003, pp. 301–305.

[57] Li, X., and Y. Mao, "A Survey on Topology Control in Wireless Sensor Networks." In *Proceedings of 10th International Conference on Control, Automation, Robotics and Vision, 2008*, Hanoi, Vietnam, December 17–20, 2008.

[58] Hollick, M., et al., "A Survey on Dependable Routing in Sensor Networks, Ad Hoc Networks, and Cellular Networks." In *Proceedings of the 30th Euromicro Conference*, August 31–September 3, 2004.

[59] Gajbhiye, P., and A. Mahajan, "A Survey of Architecture and Node Deployment in Wireless Sensor Network." In *Proceedings of First International Conference on the Applications of Digital Information and Web Technologies*, Ostrava, Czech Republic, August 4–6, 2008.

[60] Yang, Y., and F. Lambert, et al., "A Survey on Technologies for Implementing Sensor Networks for Power Delivery Systems." In *Proceedings of IEEE Power Engineering Society General Meeting*, Tampa, FL, June 24–28, 2007.

[61] Hadim, S., N. Mohammed, "Middleware for Wireless Sensor Networks: A Survey." In *Proceedings of the First International Conference on Communication System Software and Middleware*, New Delhi, India, 2006, pp. 1–7.

[62] Akylidiz, I. F., et al., "A Survey on Sensor Networks," *IEEE Communications Magazine*, August 2002, pp 102–114.

[63] Strassberg, D., "Simple Networks Will Free Many Sensors From Wires," *EDN*, April 13, 2006.

[64] Reynolds, F., "Whither Bluetooth?", *IEEE Pervasive Computing*, July–Sept 2008, Vol 7, Issue 3.

[65] Johnson, D., "Hardware and Software Implications of Creating Bluetooth Scatternet Devices," 7th AFRICON International Conference in Africa, September 17, 2004.

[66] Hasegawa, M. K., et al., "An Architecture, Topology and Performance of the Multihop Bluetooth Network." In *Proceedings of the 59th IEEE Vehicular Technology Conference*, May 17–19, 2004.

[67] Castano, G., and F. J. Garcia-Reinoso, "Survivable Bluetooth Location Network." In *Proceedings of the IEEE International Conference on Communications*, May 11–15, 2003.

[68] Padgette, J. D., "Bluetooth Security in the DoD," MILCOM 2009, Boston, MA, October 18–21, 2009.

[69] McDermott-Wells, P., "Bluetooth Scatternet Models," *IEEE Potentials*, December 2004–January 2005, Vol. 23, No. 5.

[70] McDermott-Wells, P. "What Is Bluetooth?," *IEEE Potentials*, December 2004/January 2005.

[71] Chatschik, B., "An Overview of the Bluetooth Wireless Technology," *IEEE Communications Magazine*, December 2001.

2
Fundamentals of Wireless Mobile Ad Hoc Networks

2.1 Introduction

An ad hoc network is an independent, self-governing network consisting of a large number of self-configuring, autonomous, and arbitrarily moving nodes with varying velocity. Nodes can be anything ranging from high-power computing devices like laptops to low-power handheld devices like PDAs. There are many characteristics that differentiate ad hoc networks from cellular networks. A cellular network consists of a base station or access points, but ad hoc networks are infrastructureless and are networked when the need arises. Usually, single hop communication takes place in a cellular network between the user and base station, whereas in ad hoc networks the communication takes place in a multihop mode due to the dynamically varying topology and high density of mobile nodes [1–3].

Mobile ad hoc network devices are battery based and hence routing protocols should be designed and optimized to conserve energy. Also, the transmission range of ad hoc networks is limited and bandwidth constrained. In ad hoc networks there may be no direct communication between all the nodes. Such a scenario is depicted in Figure 2.1. Node A and node C cannot communicate directly, as their communication range is out of sight. But node B can communicate with Node A and node C. Thus, multihop communication takes place between node A and node C through node B [4–6].

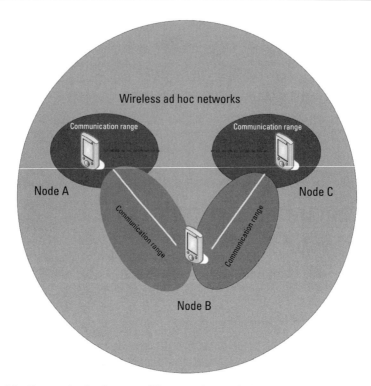

Figure 2.1 Communication between different nodes in wireless ad hoc networks.

2.2 Literature Background

In [7] the authors have compared and studied the performance effects of MAC protocols over various routing protocols in an ad hoc network. The MAC protocols considered are carrier sense multiple access (CSMA), multple access with collision avoidance (MACA), floor acquisition multiple access (FAMA), and IEEE 802.11 DCF, while the routing protocols chosen are ad hoc on-demand distance vector (AODV), wireless routing protocol (WRP), and fisheye state routing (FSR). The performance of WRP and FSR remains consistent across all the MAC protocols. But the performance of AODV changes with different MAC protocols. WRP has the highest packet delivery. CSMA with request to send/clear to send (RTS/CTS) mechanism improves the performance of all the routing protocols. The IEEE 802.11 with a collision avoidance scheme increases the packet delivery among all the routing protocols. The overhead of WRP increases with mobility. WRP has the highest normalized routing load (NRL) than FSR and AODV. The best routing protocol to use with IEEE 802.11 is AODV.

Energy can be conserved by considering a topology control specific protocol as in [8]. Initially, existing routing protocols like AODV, destination-sequenced distance-vecotr rating (DSDV), dynamic source routing (DSR), and FSR are compared to find the amount of energy consumed by each of them in idle and nonidle modes. The results show that reactive protocols like AODV and DSR consume less energy when compared to a proactive protocol like DSDV. Geographic adaptive fidelity protocols group together various redundant nodes found in the network and are grouped together based on their location information. The location information is obtained through GPS. Once the redundant nodes are recognized, then their radios are turned off resulting in saving of energy.

AODV and OLSR routing protocols are compared in an indoor environment [9]. The test bed consists of laptops running Linux operating system equipped with IEEE 802.11b cards. Metrics considered are overhead and delay. The authors show that a proactive routing protocol like OLSR performs better than a reactive protocol like AODV in an indoor environment with fewer hops.

A broadcasting protocol that is adaptive to the distance is proposed in [10]. This protocol considers flooding as the basic broadcasting mechanism. But broadcasting of packets is done based on the distance of the transmitting node in relation to its previous node. A set of nodes is chosen to rebroadcast, and only the outmost neighboring nodes are allowed to rebroadcast. These outmost neighboring nodes are identified based on the signal strength from the transmitting node. The signal strength gives the distance of each of the neighboring nodes from the transmitting node. Less delay and high efficiency is achieved by this protocol at varying mobility and high node density.

Performance comparison of on-demand multicast routing protocol (ODMRP) and ad hoc demand driven multicast routing (ADMR) protocol is carried out in [11] using different mobility models like random waypoint mobility model, Manhattan mobility model, and random drunken mobility model. The metrics considered are throughput, delay, and overhead, and these metrics are mapped against speed. Results show that ODMRP has highest throughput than ADMR at lower speed, but the throughput of ODMRP decreases with the increase in speed. ADMR has higher delay and overhead than ODMRP across all the mobility models.

Two multicast routing protocols, namely, serial multiple disjoint trees multicast routing protocol (MDTMR) and adaptive core multicast routing (ACMR) protocol, are evaluated using the Glomosim simulation library [12]. Packet delivery ratio and end-to-end delay are the metrics considered for simulation. ACMR has higher packet delivery than the MDTMR, but the packet delivery for both the protocols reduces dramatically once the mobility increases above 5 m/s. ACMR has less delay when compared to MDTMR.

Internet connectivity in mobile ad hoc networks is provided in [13] using a modified version of AODV. Proactive, reactive, and hybrid are the three methods used to discover the gateways for mobile ad hoc networks. The gateway itself initiates the route discovery process in the proactive gateway discover scenario. This is done by using gateway advertisements (GWADV), which are broadcasted periodically. One of the disadvantages of proactive gateway discovery mechanism is its flooding nature. In reactive mechanism the mobile nodes itself initiates the gateway discovery. This is done by sending the route request message to all the multicast addresses in the ad hoc network. The disadvantage of this mechanism is that a large amount of information gets accumulated at the intermediate nodes, and it takes a large amount of resources to forward this information from one node to another node. To overcome the disadvantages of proactive and reactive approaches, the hybrid approach is used. The metrics used to compare these three approaches are packet delivery ratio, end-to-end delay, and overload. Proactive and hybrid approaches have high packet delivery ratio and less end-to-end delay when compared to the reactive approach. The overhead of proactive and hybrid approaches is greater than the reactive approach. This is due to the generation of packets irrespective of whether or not the routes are needed by the proactive and hybrid approaches.

A security-aware routing protocol is proposed in [14]. The authors have adopted AODV as the base routing protocol and have modified it to suit the security needs of ad hoc networks. This modified AODV routing protocol with security-enabled features is called as secured AODV (SAODV). The routes are established using the route request and route reply packets, as in the basic AODV routing protocol. But security specific changes are incorporated into the route request packet itself so that a certain degree of security is associated with each of the forwarding nodes while establishing the routes from source to destination. In SAODV, routes are not established on the basis of least hop count but on the amount of security that is guaranteed between the nodes. Trust level is incorporated in route discovery for both outsider and insider attacks. To prevent outsider attacks the SAODV supports the unique encryption and decryption key. This security measure is incorporated at each trust level, thus preventing other users or nodes from reading it. Insider attacks can be prevented by using biometric techniques. SAODV provides resistance against interruption, interception, and subversion of messages.

2.3 Applications of Mobile Ad Hoc Networks

The applicability of mobile ad hoc networks can be found in various areas (Figure 2.2). Ad hoc networks are used to provide communication for military in the battlefields. Ad hoc networks can also be employed for rescue and disaster relief works. Ad hoc networks are deployed at conferences, meeting room, airports, and convention centers. It can also be used in home and office setups.

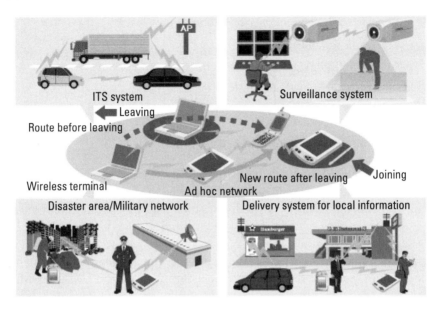

Figure 2.2 Applicability of an ad hoc network. (*Source:* Yokohama Research Laboratory, Hitachi, Ltd. Reprinted with permission.)

2.3.1 Characteristics of MANETs

Characteristics of MANETs include the following (Table 2.1):

- Multihop routing;
- Autonomous and infrastructureless;

Table 2.1
MANETs Applications

Sl. No.	Applications	Description/Services
1	Home and enterprise	Home/office WLAN, PAN
2	Emergency services	Search and rescue operations, disaster recovery, early retrieval, and transmission of patient data
3	Commercial environments	E-commerce, business, vehicular services
4	Educational applications	Virtual classrooms or conference rooms; ad hoc communication during conferences, meetings, or lectures
5	Entertainment	Multi-user games, robotics pets, outdoor Internet access
6	Sensor networks	Home applications with smart sensors, environmental applications including tracking the movement of animals, tracking data highly correlated in time and space
7	Tactical networks	Military communication, operations, automated battlefields
8	Location aware services	Follow-on services (e.g., automatic call forwarding information services), push (e.g., advertise location specific service), and pull (e.g., location dependent travel guide)

- Dynamically changing network topologies;
- Network scalability;
- Variation in link and node capabilities;
- Energy-constrained operation.

2.4　Medium Access Control Layer

The MAC layer performs various functionalities like error detection and resolving contention among various nodes while sharing the same channel in a decentralized manner. Efficient bandwidth utilization and throughput improvement are the main criteria while designing MAC protocols. MAC protocols can be classified into contention free and contention based protocols. In contention free MAC protocols, data transmission takes place without any contention for the medium. Some of the protocols under this scheme are TDMA, CDMA, and FDMA. Contention free MAC protocols are used in applications that are sensitive to delay and where bandwidth is a constraint. With contention based MAC protocols, if a packet has to be transmitted then the transmitting node has to contend with neighboring nodes to access the channel. Infrequent data transmission takes place in contention based schemes [15–17].

IEEE 802.11 is the most popular MAC protocol to be used in ad hoc network. Various problems like hidden node problem, exposed node problem, and unfairness are commonly found when IEEE 802.11 is used in a multihop network like ad hoc network.

- *Hidden node problem:* consider three nodes: node A, node B, and node C. Node A and node C are within the communication range, while node B and node C are also within the communication range. But node A and node B are out of range from each other. When node A and node C are communicating with each other, then a situation may arise wherein node B tries to communicate with node C. As node C is out of range from node A, it does not know that node A is communicating with node B. This result in a collision of packets. Various solutions have been proposed to prevent the hidden node problem like RTS/CTS mechanism and carrier sense mechanism [18].

- *Exposed node problem:* Consider four nodes: node A, node B, node C, and node D. Node A can communicate only with node B and in the same way node D can communicate with node C. Also node B and node C are assumed within the communication range of each other. Imagine both node B and node C has packets to transmit. If node B and node A are communicating with each other, then node C won't transmit any packets to node D even when it is in the same communication range of node D. This happens because node C thinks that it will interfere with the neighboring node B due to carrier sense [19].

- *Unfairness:* The sensing range is more than the actual transmission in an IEEE 802.11 ad hoc environment. Thus further increases the exposed node and hidden node problems in a multihop network like an ad hoc network. If the communication between any two nodes is asymmetric then it may result in unfairness in the network. If a node is heavily loaded with data then it continuously tries to access a channel by sending the packets again and again. This prevents nodes that are loaded lightly from accessing the channels, thereby resulting in unfair access of channels [20, 21].

An overlay MAC layer is proposed in [22]. This is a software solution to extend the capabilities of IEEE 802.11. This overlay MAC layer reduces the queuing of packets and eliminates any interference from neighboring nodes. This is achieved by reducing the buffer size and by synchronizing the clocks.

A practical study of hidden node problem of IEEE 802.11 is carried out in [23]. The default data rate in NS2 network simulator is 2 Mbps. For the experimental setup, the data rate is fixed at 11 Mbps. The results from both the simulation setup and experimental setup were compared. The authors show that transmitting data either at 2 Mbps or 11 Mbps does not make difference to the overall performance of the network.

2.5 Topology Control

Topology control is an absolute requirement in a multihop network like an ad hoc network, which consists of mobile nodes. Topology control ensures that the network maintains a certain degree of connectivity in the topology so that the overall network does not suffer performance degradation due to the dynamic movements of nodes. Topology refers to the availability of the routes between various nodes. The establishment of routes depends on various factors like location of the nodes, availability of energy in each node, transmission range, and direction of antennas [24]. Mismanagement of network topology can lead to various problems as specified in [25, 26]. A network with fewer nodes in a topology may result in network partition, less throughput, and higher delay in the network. If the network gets partitioned, then no routes are available to reach the destination. Dense node topology increases the number of routes available in the network, which may result in collision of packets. This again leads to more utilization of energy.

Topology control in ad hoc networks is carried out through power control and through hierarchical organization. In the power control nodes, the amount of energy required for transmission and receiving of packets is done on a per-node basis. In a hierarchical organization, various clusters are formed to manage topology. Topology based on clustering mechanism leads to fewer links but maintains high connectivity, thereby conserving the amount of energy spent in establishing routes in the network [27].

In literature a majority of the papers argue that there should be a minimum transmission range to maintain a topology where all the nodes are connected. But in [28] the authors argue that this minimum transmission range connected topology may lead to reduced performance in ad hoc networks. Through experiments it is proved that high throughput is achieved for fewer nodes in a network at minimum transmission range. When the nodes are increased, the throughput reduces even when the transmission range is kept at minimum.

Interference among neighboring nodes can be reduced by appropriately configuring the topology of a network. A critical neighbor scheme proposed in [26] improves the overall network performance by adaptively varying the transmission power of the nodes in the network based on route and traffic demands. This critical neighbor scheme is based on measuring the critical range, then estimating and adjusting the ideal power. Simulation results show that a routing protocol with a critical neighbor scheme has higher throughput and higher packet delivery.

2.6 Routing Protocols

Ad hoc networks consist of huge numbers of mobile nodes. A packet has to travel in multihop to reach a destination node. Also, there might be path breaks due to mobility of the nodes, which requires finding new path to the destination. Thus, reestablishment of paths has its effect on scarce resources like energy and bandwidth. This may also induce delay, thereby significantly degrading the network performance. Therefore, routing should be designed in such a way that the routes are established with minimum exchange of messages across the network [29, 30]. Routing protocols can be classified into proactive and reactive routing protocols. With proactive routing protocols, every node maintains information regarding every other node in the network to establish the routes, while reactive routing protocols routes are established only when there is a need for paths between the source node and destination node. The main challenges faced in an ad hoc network are topology, bandwidth, power, and security [31]. In [32] the authors have identified five factors that affect the performance of ad hoc networks. They are (a) node speed, (b) pause time, (c) network size, (d) number of traffic sources, and (e) routing protocol. The various characteristics while designing routing protocols, classification of routing protocols, and experimental evaluation of various routing protocols under diversified scenarios are conducted in great detail in Chapter 3.

2.7 Broadcasting

Broadcasting is a technique in which a source node sends messages to all its neighboring nodes in order to establish a path to the destination node. Broadcasting is a preferred choice in a majority of the reactive routing protocols like AODV, DSR, DYMO, and AOMDV. The various problems associated with broadcasting are spontaneous broadcasting and unreliability. Broadcasting can be spontaneous in nature because any node in the network can be selected to broadcast a packet at any instant of time. This results in less preparation for optimizing the broadcasted packets. Broadcasting can be unreliable in ad hoc networks. The nodes in the ad hoc environment are mobile and the neighboring nodes may move out of the transmission range when the transmitting node is about to transmit a packet [33]. Broadcast based protocols can be classified into simple flooding, probability based methods, area based methods, and neighbor based methods [34–36]. In simple flooding the packets generated from the source node are broadcasted across all the nodes in the network. In probabilistic based method a certain degree of probability is assigned to each of the nodes to broadcast the packets based on the network topology. In area based broadcasting rebroadcast of the packets is done if the packets are able to reach or cover a certain amount of area. The neighbor knowledge broadcasting technique uses hello packets to keep a tab on its neighboring nodes.

Another classification of broadcasting protocols is done in [37–39]. Broadcasting is classified as optimized broadcast with fixed radius and optimized broadcast with variable radius. The various techniques under optimized broadcast with fixed transmission range are heuristic based broadcasting, neighbor coverage based broadcasting, dominating sets based broadcasting, combination of multipoint relay and dominating set based broadcasting, hexagonal and dominating set broadcasting, cluster based broadcasting, and resource-aware broadcasting.

Under optimized broadcasting with varying transmission range, broadcasting methods are relative neighborhood graph, minimum spanning tree graph, neighbor-aware adaptive power broadcasting, and incremental broadcasting. Based on the present conditions of the ad hoc networks, the tuning of concerned parameters is done in heuristic based broadcasting. Information regarding neighboring nodes is obtained by periodically broadcasting the beacon message in the neighbor coverage based broadcast method. Connected dominating set broadcast technique is used to overcome the broadcast storm problem. In multipoint relay broadcasting algorithm, a relay node should collect information from the previous node. Hexagonal and dominating set based broadcasting considers the hexagonal tiling of a plane to broadcast the packets. In cluster based broadcasting, various nodes are grouped into a cluster based on some criteria, and a cluster head is selected based on the availability of resources for that node. Communication takes place between nodes in different clusters through the cluster heads and some special nodes called gateway nodes that may be common among the two communicating clusters. Resources-aware broadcasting makes use of available resources like bandwidth and energy in an optimized way to broadcast the packets. Consider two nodes that are connected by an edge in the lune of the graph. If this lune does not contain any other nodes in the network then a relative neighbor graph (RNG) is formed. To overcome the broadcast storm problem, a relay neighborhood graph can be used where the neighboring nodes of the RNG receive the packets. The minimum spanning tree (MST) broadcasting method has its own advantages and disadvantages. The paths generated in MST are optimal when compared to RNG paths. But this approach does not check for fault tolerance. The MST is constructed based on a common transmission range. A common transmission range is the longest communication range used to maintain connectivity among all the nodes. The common transmission range is decided based on the longest edge in MST. In the neighbor-aware adaptive power broadcasting technique, a node first determines the nodes that have not received the same packets as it received from another node. This node then forwards a packet to its neighboring node that still has not received the same packet if the receiving node is nearer to the transmitting node. By intelligently adapting the power for the transmission range, only those neighboring nodes that are closer to transmitting nodes receive the broadcasted packets.

The broadcast storm problem is discussed in [39]. The broadcast storm problem is caused due to flooding. The broadcast storm problem may result in redundant broadcasts, contention, and collision in the network. Various solutions have been proposed to counter this problem, like the probabilistic scheme, counter based scheme, distance based scheme, location based scheme, and cluster based scheme. A reliable broadcast algorithm is proposed in [40]. In this algorithm, a scheduling technique is employed that takes the relative position of the nodes into consideration based upon the source node. If the distance between the transmitting node and the destination node is too far, then transmitting node is allowed to transmit with immediate effect. The amount of messages that can be broadcasted by a transmitting node is minimized by limiting transmission unless there is a new node to which it has to transmit; otherwise, transmission is canceled. So this contention free broadcasting algorithm achieves reliability in O(Dlog n) time slots where D is the diameter of the network and n is the number of nodes.

2.8 Multicasting

Multicasting is a technique in which the packets are delivered to a group of nodes having the same destination address instead of all the nodes in the network. Some of the advantages of multicasting are reduced energy consumption, efficient use of bandwidth, and reduced transmission cost [41, 42]. In multicasting, the routing protocols come under various categories like (a) reactive, flooding, and proactive; (b) tree and mesh based protocols; (c) location based; (d) quality of service based protocols; (e) energy efficiency; and (f) network coding [43–45]. Tree based protocols involve establishing single paths. In mesh based routing protocols, a large number of redundant paths are established. In location based routing protocols, the information regarding the location of nodes is obtained through GPS. The quality of service based protocols is used to serve applications of various kinds like audio and video [46, 47].

A study of overlay multicasting routing protocols is done in [48]. Conserving energy is the main requirement of ad hoc networks. The multicast routing protocols should take this parameter into consideration. A node supported by network coding receives the packets from all the input links, and the received information is encoded before rerouting it to all the output links. In general an overlay multicasting routing protocol involves the construction of a virtual tunnel among multicast group nodes. This virtual tunnel supports multicast routing. This virtual overlay technique reduces the overhead in the network.

A new protocol called source routing based multicast protocol (SRMP) is proposed in [49]. This SRMP protocol is compared with ODMRP and ADMR multicast routing protocol. SRMP shows better packet delivery and less overhead then ODMRP and ADMR.

Evaluation of multicast routing protocols based on flooding and associativity is done in [50]. ODMRP is the flooding based multicast routing protocol, and ABAM is the associativity based multicast routing protocol.

Various metrics are evaluated against a fraction of moving nodes. At lower speeds, ABAM has higher throughput than ODMRP but decreases with the increase in mobility speed. ODMRP, on the other hand, maintains a constant performance even when the mobility speed is increased. ODMRP shows higher control overhead then ABAM, and the difference is huge. This is due to the flooding nature of ODMRP. Also the end-to-end delay of ODMRP is higher when compared to ABAM.

2.9 Internet Connectivity for Mobile Ad Hoc Networks

Providing Internet connectivity to mobile nodes in an ad hoc network is a scintillating idea. Gateways are special nodes that act as a bridge and are used by the mobile nodes in the ad hoc network to connect to the fixed hosts in an IP network [51, 52]. Some of the challenges for connecting ad hoc networks to a fixed IP network are [53–55].

- *The hierarchical IP addresses:* Due to the mobility nature of nodes in an ad hoc network, the topology changes dynamically. This results in nodes moving from one subnet to another. Thus, the IP address for a node does not show its original location from which its IP address was assigned.
- *Dead zone:* This is another problematic area. The mobile nodes in an ad hoc network need to connect to hosts in the fixed network through a gateway. But just before joining a gateway a mobile node may lose its communication range thus getting disconnected.
- *Discovering Internet gateways:* This process is the vital part in ad hoc network–Internet connectivity mechanism. It also influences the overall performance of the network. Gateways can be discovered using proactive, reactive, and hybrid approaches.
- *Address auto configuration:* Mobile nodes that wish to communicate with the IP network have to obtain the IP addresses, and this is done by using a DHCP server.
- *Reaching a destination:* A mechanism should be incorporated within the routing protocols so that a source node wishing to communicate with a destination node should know whether the destination is present in the ad hoc network or in the fixed IP network. Due to the mobility nature of ad hoc networks, the nodes can move across different regions. This may result in the duplication of addresses and has to be avoided. Name resolution should be provided for ad hoc nodes connecting to fixed IP networks.

Even though HELLO messages can lead to large overhead and higher end-to-end delay, a new scheme proposed in [56] shows improvement over proactive and reactive methods in terms of route discovery time and handover time.

Some of the solutions proposed to overcome the Internet connectivity problems are [57] (a) route requests to establish that a connection should be initiated by the mobile nodes of the ad hoc networks and not by the hosts in the fixed network, and (b) the mobile nodes should update the gateways of their location every time the location changes the base station or moves from one subnet region to another.

2.10 Security in Mobile Ad Hoc Networks

Ad hoc networks are inherently prone to security breaches. In most of the cases, nodes in an ad hoc network are unattended and the self organizing behavior of the wireless networks can act as pot of gold to extract information from an unsuspecting user. To secure the ad hoc networks, the following properties should be considered [58]:

- *Availability:* A network should be up and running even if it comes under attack. Denial of service can affect the availability of the network at all the layers. Jamming and tampering affects the physical layer, and at the network layer routing of packets can be affected.
- *Confidentiality:* If ad hoc networks are deployed in a military environment, then every bit of information is of high importance and confidential. Networks should be secured so that confidential data does not end up in the hands of enemies.
- *Integrity:* Ad hoc networks are multihop networks, and the packets have to pass through several nodes before reaching the destination. An adversary may corrupt the data before the packets reach the destination. Also, the data received at the destination should be the same as it was at the originated node.
- *Authentication:* Every node communicating with every other node should keep information about the neighboring node with which it intends to communicate. Without authentication, an adversary may introduce a malicious node in the network that may communicate with the genuine nodes and thus enable the adversary to lay her hand on important information.
- *Nonrepudiation:* In nonrepudiation, an account is maintained for every node from which a message gets originated. Even if a node gets compromised, then that node can be separated by maintaining this information.

In Table 2.2, the various types of attacks that can happen at different layers in mobile ad hoc networks are summarized, and countermeasures for these attacks are discussed [59–65].

Table 2.2
Type of Attack at Different Layers in MANETs and Their Countermeasures

Different Layers	Type of Attack	Countermeasure
Physical layer	Jamming	Various forms of spread-spectrum technique, like frequency hopping, are used. If the hopping sequence is done very fast, then the jammer cannot interfere with the communication.
	Tampering	Self-destruction—whenever somebody accesses the sensor nodes physically, then the nodes vaporize their memory contents.
		Hiding nodes.
Data link layer	Collision induction	Error correcting codes can be incorporated in the data packets to defend against collision.
		Collision detection.
		Collision-free MAC.
	Battery exhaustion	Rate limitation—if a node transmits more than a threshold number of times, then the node should be put in sleep mode.
		Streamlined protocols.
Network layer	Black hole attack	Need authentication mechanism.
	Misdirection and Internet smurf attack	If a node's network link is getting flooded without any useful information, then the victim node can be scheduled into sleep mode for some time.
	Rushing attack	Detect secure neighbor.
	HELLO Flood attack	Check bidirectional links whenever selecting a path.
	Spoofing	Efficient encryption and authentication techniques.
Transport layers	Flooding	Limit number of connections from a particular node.
	Desynchronization	Authenticate all messages including header fields.
Application layer	Subversion and malicious nodes	Malicious node detection and isolation.

References

[1] Li, V. O. K., and Z. Lu, "Ad Hoc Network Routing." In *Proceedings of the 2004 IEEE International Conference on Networking, Sensing and Control*, Taipei, Taiwan, March 21–23, 2004.

[2] Dow, C. R., and P. J. Lin, et al. "A Study of Recent Research Trends and Experimental Guidelines in Mobile Ad Hoc Networks." In *Proceedings of the 19th International Conference on Advanced Information Networking and Applications, 2005*, AINA 2005, Vol. 1, pp. 72–77.

[3] Frodigh, M., P. Johansson, and P. Larsson, "Wireless Ad Hoc Networking: The Art of Networking Without Network," *Ericsson Review*, No 4, 2000, pp. 248–263.

[4] Sun, J.-Z., "Mobile Ad Hoc Neworking: An Essential Technology for Pervasive Computing." In Proceedings of 2001 International Conference on Info-tech and Info-net, Beijing, China October 29 – November 1, 2001.

[5] Ramnathan, R., and J. Redi, "A Brief Overview of Ad Hoc Networks: Challenges and Directions," IEE Communications Magazine, 50th Anniversary Commemorative Issue, May 2002, pp 20-22.

[6] Subbarao, M. W., "Ad Hoc Networking Critical Features and Performance Metrics," Wireless Communications Technology Group, NIST, October 7, 1999.

[7] Royer, E. M., S.-J. Lee, and C. E. Perkins, "The Effects of Mac Protocols on Ad Hoc Network Communication." In proceedings of IEEE WCNC 2000, Chicago, IL, September 2000.

[8] Xu, Y., D. Estrin et al., "Topology Control Protocols to Conserve Energy in Wireless Ad Hoc Networks," Technical Report 6, University of California, Los Angeles, Center for Embedded Networked Computing, January, 2003.

[9] Borgia, E., "Experimental Evaluation of Ad Hoc Routing Protocols," Proceedings of the 3rd IEEE International Conference on Pervasive Computig and Communications Workshops (PerCome 2005 Workshops), 8th to 12th March 2005, Kauai Island, HI, USA.

[10] Faloutsos, X. M., S. Krishnamurthy, "Distance Adaptive (DAD) Broadcasting for Ad Hoc Networks." In *Military Communications Conference*, MILCOM, 2002, pp. 878–882.

[11] Malarkodi, B., P. Gopal, and B. Venkataramani, "Performance Evaluation of Ad Hoc Networks with Different Multicast Routing Protocols and Mobility Models," International Conference on Advances in Recent Technologies in Communication and Computing, Kerala, India, October 27–28, 2009.

[12] Lee, J.-H., J.-W. Cho, et al., "Performance Comparison of Mobile Ad Hoc Multicast Routing Protocols," International Conference on Advanced Technologies for Communications, Hanoi, Vietnam, October 6–8, 2008.

[13] Hamidian, A. A., and U. Korner, et al., "Performance of Internet Access Solutions in Mobile Ad Hoc Networks," *Lecture Notes in Computer Science*, Vol. 3427, June 2004, pp. 189–201.

[14] Yi, S., P. Naldurg, and R. Kravets, "A Security-Aware Routing Protocol for Wireless Ad Hoc Networks." In *Proceedings of the 2nd ACM International Symposium on Mobile Ad Hoc Networking and Computing*, Long Beach, CA, 2001.

[15] Van den Heuvel, S., "A Survey of MAC Protocols for Ad Hoc Networks and IEEE 802.11," *Proceedings of the 4th National Conference MiSSI 2004*, Poland, pp. 23–33.

[16] Rubinstein, M. G., I. M. Moraes, and M. E. M. Capista, "A Survey on Wireless Ad Hoc Networks," IFIP, ed., Boston, MA, Vol. 211, August 2006, pp. 1–33.

[17] Kumar, S., V. S. Raghavan, and J. Deng, "Medium Access Control Protocols for Ad Hoc Wireless Networks: A Survey," *Ad Hoc Networks*, Vol. 4, 2006, pp. 326–358.

[18] Rahman, A., and P. Gurzynski, "Hidden Problems with the Hidden Node Problem." In *Proceedings of 23rd Biennial Symposium on Communications*, 2006.

[19] Jayasuriya, A., and S. Perreau, et al., "Hidden vs. Exposed Terminal Problem in Ad Hoc Networks." In *Proceedings of the Australian Telecommunications, Networks and Architecture Conference (ATNAC 2004)*, Sydney, Australia, December 2004.

[20] Xu, S., and T. Saadawi, "Does the IEEE 802.11 MAC Protocol Work Well in Multihop Wireless Ad Hoc Networs?" *IEEE Communications Magazine*, June 2001.

[21] Chaudet, C., D. Dhoutaut, and I. G. Lassous, "Performance Issues with IEEE 802.11 in Ad Hoc Networking," *IEEE Communications Magazine*, July 2005.

[22] Rao, A., and I. Stoica, "An Overlay MAC Layer for 802.11 Networks." In *Proceedings of the 3rd International Conference on Mobile Systems, Applications and Services*, Seattle, WA, 2005, pp. 135–148.

[23] Ng, P. C., S. C. Leew, K. C. Sha, and W. T. To, "Experimental Study of Hidden Node Problem in IEEE 802.11 Wireless Networks," Poster Presentation, SIGCOMM 2005, PA.

[24] Bao, L., and J. J. Garcia-Luna-Aceves, "Topology Management in Ad Hoc Networks," MobiHoc'03, Annapolis, MD, June 1–3.

[25] Ramnathan, R., and R. Rosales-Hain, "Topology Control of Multihop Wireless Networks Using Transmit Power Adjustment." In *Proceedings of IEEE Infocom*, March 2000.

[26] Tan, H. X., and W. K. G. Seah, "Dynamic Topology Control to Reduce Interference in MANETs." In *Proceedings of the 2nd International Conference on Mobile Computing and Ubiquitous Networking*, Osaka University Convention Centre, Osaka, Japan, April 13–15, 2005.

[27] Venkatesan, S., and C. D. Young, "A Distributed Topology Control Algorithm for MANETS," *Proceedings of Military Communications Conference (MILCOM) 2005*, Atlantic City, NJ, October 2005.

[28] Park, S.-J., and R. Sivakumar, "Adaptive Topology Control for Wireless Ad Hoc Networks," MobiHoc'03, Annapolis, MD, June 1–3, 2003.

[29] Conti, M., and S. Giordano, "Mobile Ad Hoc Networking." In *Proceedings of the 34th Hawaii IEEE International Conference on Systems Sciences*, 2001.

[30] Feenay, L. M., and B. Ahlgren, et al., "Spontaneous Networking: An Application Oriented Approach to Ad Hoc Networking," *IEEE Communications Magazine*, June 2001.

[31] Latiff, L. A., and N. Fisal, "Routing Protocols in Wireless Mobile Ad Hoc Network—A Review." In *Proceedings of the 9th Asia-Pacific Conference on Communications (APCC2003)*, Penang, Malaysia, September 2003.

[32] Perkins, D. D., H. D.Hughes and C. B. Owen, "Factors Affecting the Performance of Ad Hoc Networks." In *Proc. of the IEEE Int. Conf. on Communications (ICC 2002)*, New York City, April 28–May 2, 2002, pp. 2048–2052.

[33] Williams, B., and T. Camp, "Comparison of Broadcasting Techniques for Mobile Ad Hoc Networks," MOBIHOC'02, Laussanne, Switzerland, June 9–11, 2002.

[34] Vollset, E., and P. Ezhilchelvan, "A Survey of Reliable Broadcast Protocol for Mobile Ad Hoc Networks," Technical Report, School of Computer Science, Newcastle University.

[35] Contractor, J., "Survey: Broadcasting in Wireless Sensor Network, Ad Hoc Network and Delay Tolerant Network," CIS Department, Temple University.

[36] Lipman, J., H. Liu, and I. Stojmenovic, "Broadcast in Ad Hoc Networks," *Guide to Wireless Ad Hoc Networks, Computer Communications and Networks,* London: Springer-Verlag, 2009.

[37] Ingerlrest, F., and I. Stojmenovic, "Energy Efficient Broadcasting in Wireless Mobile Ad Hoc Networks," *Resource Management in Wireless Networking*, 2004.

[38] Ingelrest, F., and I. Stojmenovic, "Routing and Broadcasting in Hybrid Ad Hoc Networks," IRCICA/LIFL, Univ. Lille 1, INRIA Futurs, France.

[39] Ni, S-Y., Y-C. Tseng, et al., "The Broadcast Storm Problem in Mobile Ad Hoc Networks," MOBICOM, Seattle, Washington, 1999.

[40] Mohsin, M., and D. Cavin, et al., "Reliable Broadcast in Wireless Mobile Ad Hoc Networks." In *Proceedings of the 39th Annual Hawaii International Conference on System Sciences,* January 4–7, 2006, p. 233.1.

[41] Kaur, J., and C. Li, "Simulation and Analysis of Multicast Protocols in Mobile Ad Hoc Networks Using NS2." In *Proceedings of IEEE NECEC 2007,* St.John's, Newfoundland, Canada.

[42] Obraczka, K., and G. Tsudik, "Multicast Routing Issues in Ad Hoc Networks," IEEE International Conference on Universal Personal Communication (ICUPC'98), October 1998.

[43] Junhai, L., Y. Danxia, et al., "A Survey of Multicast Routing Protocols for Mobile Ad Hoc Networks," IEEE Communications and Tutorial, 2009.

[44] Junhai, L., X. Liu, and Y. Danxia, "Research on Multicast Routing Protocols for Mobile Ad Hoc Networks," *Computer Networks,* Vol. 52, No. 5, 2008, pp. 988–997.

[45] Ali, M., et al., "A Survey of Multicasting Routing Protocols for Ad Hoc Wireless Networks," *Minufiya Journal of Electronic Engineering Research,* Vol. 17, No. 2, July 2007.

[46] Striegel, A., and G. Manimaran, "A Survey of QOS Multicasting Issues," *IEEE Communications Magazine,* Vol. 40, No. 6, 2002, pp. 82–87.

[47] Wang, B., and J. C. Hou, "Multicast Routing and Its QoS Extension: Problems, Algirthms and Protocols," *IEEE Networks,* Vol. 14, No. 1, 2000, pp. 22–36.

[48] Bing, S., Y. Haiyang, et al., "Research on Overlay Multicast Routing Protocols for Mobile Ad Hoc Networks." In *Proceedings of the 5th International Conference on Wireless Communications, Networking, and Mobile Computing,* Beijing, China, September 24–26, 2009, pp. 3030–3033.

[49] Moustafa, H., and H. Labiod, "A Performance Comparison of Multicast Routing Protocols in Ad Hoc Networks." In *Proceedings of the 14th IEEE 2003 International Symposium on Personal, Indoor and Mobile Radio Communications,* Beijing, China, September 7–10, 2003.

[50] Toh, C.-K., and S. Bunchua, "Performance Evaluation of Flooding-Based and Associativity-Based Ad Hoc Mobile Multicast Routing Protocols," IEEE Wireless Communications and Networking Conference, Vol. 3, 2000, pp. 1274–1279.

[51] Bahety, V., and R. Pendse, "Bridging the Generation Gap," *IEEE Communications*, October/November 2005.

[52] Corsor, S., and J. P. Macker, "Internet-Based Mobile Ad Hoc Networking," *IEEE Internet Computing*, July/August 1999.

[53] Sun, Y., E. M. Royer, and C. E. Perkins, "Internet Connectivity for Ad Hoc Mobile Networks," *International Journal of Wireless Information Networks*, Vol. 9, No. 2, April 2002.

[54] Ros, F. J., P. M. Ruiz, et al., "Performance Evaluation of Interconnection Mechanisms for Ad Hoc Networks Across Mobility Models," *Journal of Networks*, Vol. 1, No. 2, June 2006.

[55] Ruiz, P. M., and F. J. Ros, et al., "Internet Connectivity for Mobile Ad Hoc Networks: Solutions and Challenges," *IEEE Communications Magazine*, October 2005.

[56] Matthias, R., V. Rakocevic, et al., "Performance Comparison of Gateway Discovery Algorithms in Ad Hoc Networks with Mobile Nodes," School of Engineering and Mathematical Sciences, City University, London, September 2005.

[57] Michalk, M., and T. Braun, "Common Gateway Architecture for Mobile Ad Hoc Networks." In *Proceedings of the Second Annual Conference on Wireless On-Demand Network Systems and Services*, January 19–21, 2005, pp. 70-75.

[58] Zhou, L., and Z. J. Haas, "Securing Ad Hoc Networks," *IEEE Network Magazine*, Vol. 13, No.6, November/December 1999.

[59] Hubaux, J.-P., and L. Buttyan, et al., "The Quest for Security in Mobile Ad Hoc Networks." In *Proceedings of the 2nd ACM International Symposium on Mobile Ad Hoc Networking & Computing*, Long Beach, CA, October 4–5, 2001.

[60] Kong, J., X. Hong, and M. Gerla, "A New Set of Passive Routing Attacks in Mobile Ad Hoc Networks," IEEE Military Communications Conference, 2003.

[61] Karpijoki, V., "Security in Ad Hoc Networks," http://www.hut.fi/~vkarpijo/netsec00/netsec00_manet_sec.ps.

[62] Gupte, S., and M. Singhal, "Secure Routing in Mobile Wireless ad Hoc Networks," *Ad Hoc Networks*, Vol. 1, No. 1, 2003, pp. 151–174.

[63] Lundberg, J., "Routing Security in Ad Hoc Networks," Tech. Rep. Tik-110.501, Helsinki University of Technology, 2000.

[64] Zapata, M. G., and N. Asokan, "Securing Ad Hoc Routing Protocols." In *Proceedings of the 3rd ACM Workshop on Wireless Security*, Atlanta, GA, September 28, 2002, pp. 1–10.

[65] Yau, P.-W., and C. J. Mitchell, "Security Vulnerabilities in Ad Hoc Networks." In *Proceedings of the 7th International Symposium on Communication Theory and Applications*, Ambleside, Lake District, UK, 2003.

3

Scenario Based Performance Analysis of Various Routing Protocols in MANETs

3.1 Introduction

Ad hoc networks are a specialized class of networks comprising of hundreds of nodes that are able to communicate among each other in a dynamic arbitrary manner. Ubiquitous computing is intertwined with the ad hoc networking. In this type of environment the devices make use of the resources provided in their current setup. Also the mobile nodes are crippled with the amount of power used in mobility and for sending and receiving the messages. Infrastructureless setup of ad hoc networks can be used in military deployment and emergency operations.

As ad hoc networks are wireless, they have no fixed infrastructure. Nodes in ad hoc networks are mobile in nature. If any two nodes are out of range, then connectivity is established by hopping through various intermediate nodes. If any two nodes are within the transmission range of each other, then the connectivity is established in a peer-to-peer manner.

Mobility models play a very important role in simulating the routing protocols. These mobility models should reverberate with real-life scenarios to obtain accurate results. The nodes in a mobile ad hoc network are inherently dyanmic; this results in frequent changes in the network technology. The change in topology is controlled by the underlying mobility models in a simulated result. It is entirely desirable for these mobility models to consider the real-world traces so that when these routing protocols are deployed in real-life systems, they do provide the same effects they provide when simulated.

In this chapter we study the comparison of six different routing protocols, namely, ad hoc on-demand multipath distance vector (AOMDV), dynamic MANET on-demand (DYMO), fisheye state routing protocol (FSR), location-aided routing protocol (LAR), optimized link state routing protocol (OLSR), and temporally ordered routing algorithm (TORA) using community-based mobility model and SMS mobility model. Of these LAR, AOMDV, and TORA routing protocols are analyzed with community mobility model, and OLSR, DYMO and FSR protocols are analyzed using SMS mobility model. The reason for bifurcation of routing protocols is to see whether these routing protocols were able to take financial advantage of the mobility model on which they were being applied. This allows us to analyze the application of routing protocols in a network scenario for a particular application, like identifying the location of the moving nodes. The community-based mobility model and SMS mobility model are modern-day mobility models with close resemblance to real-world vehicular or human group mobility. It is still not clear how these routing protocols behave under these mobility models. Our research tries to address various issues like effect of mobility, traffic, energy consumption, and quality of services of these routing protocols on the mentioned mobility models. The results obtained colligates with the results obtained in [1].

Also, we have compared AOMDV and OLSR routing protocols using the Levy walk mobility model and Gauss-Markov mobility model. OLSR is a proactive, table-driven, link state routing protocol, while AOMDV is a reactive routing protocol. Besides eyeing the details of comparing a proactive routing protocol with reactive routing protocol, we try to address the circumstantial effect of these mobility models on the routing protocols. Mobility models should symbolize the precise movement of nodes. A thorough investigation of these routing protocols over mobility models that reflect real-world scenarios are very important as they affect the overall performance of the network. Part of the results obtained is in conjugation with [2], thus providing a thorough reference for other research works.

The main contribution of this chapter is as follows:

- An inviolable effort to study the performance of six routing protocols (AOMDV, DYMO, FSR, LAR, OLSR, and TORA) over a community-based mobility model and SMS mobility model;
- A solid effort has been made to study the performance of AOMDV and OLSR routing protocols over a Levy walk mobility model and Gauss-Markov mobility model.

To the best of our knowledge, no work has been reported that compares and studies the performance of all six of these routing protocols with a community mobility model, SMS mobility model, Levy walk mobility model, and Gauss-Markov mobility model.

3.2 Literature Background

In [3] the routing protocols are compared in MANET and sensor network scenarios. In MANET scenario protocols like ad hoc on-demand distance vector (AODV), destination-sequenced distance-vector routing (DSDV), and TORA are compared. In the sensor scenario, AODV, DSDV, TORA, and low-energy adaptive clustering hierarchy (LEACH) protocols are considered. The authors came to conclusion that AODV is the best routing protocol in MANET with less overhead and higher packet delivery. In a sensor network environment, the LEACH protocol has improved performance over AODV. It is also shown that that TORA is the worst protocol in both the scenarios.

Five different mobility models—random waypoint, random direction, Gauss Markov model, city section and Manhattan mobility model—are compared in [4]. These mobility models are compared using link disjoint and node disjoint multipath routing algorithms. With lifetime as the parameter, the Gauss-Markov mobility model and random waypoint mobility model have highest lifetime at link disjoint and node disjoint multipath at a low speed. The lifetime of these mobility models decreases as the speed increases. For a multipath set size parameter, the maximum number of paths are generated for city section and random waypoint at both low speed and high speed.

A performance analysis of AODV and OLSR traffic load and node density is done in [5]. The parameters considered are packet delivery and the number of collisions per packet. The authors show that with less mobility and more traffic load, the delivery of packet is highest. But with increase in mobility and the traffic load, the packet delivery decreases. This trend is observed for both AODV and OLSR. There is an increase in the number of collisions that occurs in the network when the number of neighboring node is increased. This is found to be severe in OLSR routing protocol.

Popular mobility models like random waypoint, random walk, and random direction mobility models are compared in [6]. Packet delivery ratio, routing overhead, and throughput are the metrics considered. These mobility models are compared by varying the number of nodes and the mobility speed. For a varying mobile density, the random waypoint has the highest packet delivery and throughput, and random direction has the highest overhead. An increase in mobility speed results in random walk having the highest routing overhead. Just as in varying node density, even with varying mobility speed, the highest packet delivery and throughput is associated with random waypoint mobility model.

A new group mobility model based on *virtual track* is proposed in [7]. This mobility model considers the military scenarios that can be applied in ad hoc networks. This mobility model is then compared with random waypoint mobility model. Through simulation, it is shown that this group mobility model based on real-world scenarios provides more realistic results than the random way point model.

Routing protocols like AODV, FSR, DSR, and TORA are compared in a city section environment [8]. In a city section model, the AODV routing protocol scores over other routing protocols and FSR comes a close second.

A very large-scale mobile ad hoc network scenario is considered in [9]. AODV, DSR, and LAR are the three routing protocols considered. In this scenario, 500 nodes are deployed over an area of 12,000 x 6,000 m. In such a large-scale network, it is shown through simulation that the packet delivery ratio of AODV is less than DSR and LAR, but AODV has less delay. DSR has a high packet delivery rate and also high delay. This is attributed to the caching strategy used in DSR. It is shown that the packet delivery rate of AODV decreases with the increase in the number of connections. The authors also compare QualNet and NS2 simulators and come to the conclusion that there is not much difference between these two simulators, and they are very much reliable when comparing the routing protocols.

Using Georgia Tech Network Simulator (GTNetS), a very large-scale network ranging from 10,000 nodes to 50,000 nodes is evaluated with AODV as the routing protocol in [10]. For this kind of setup, the packet delivery rate is around 32 percent. This is too low, as the same routing yielded more than 90 percent packet delivery for 50 nodes. Increasing the mobility speed does not seem have much effect on the packet delivery performance of AODV, but higher packet delivery was observed with increase in pause time. Increase in mobility increases the end-to-end delay in the network, but the increase in pause time results in less delay. Simulation results also show that the AODV routing protocol with thousands of nodes and lower mobility with more pause time results in less overhead in the network.

In [11] the simulation results are divided in to two parts. In the first part results are obtained by varying the number of nodes, and in the second part the results are obtained by moving the nodes in a directed trajectory so that these nodes move from an out-of-range zone to a communication zone and back again. Parameters considered are end-to-end delay, throughput, and media access delay. For varying node density, the routing protocols compared are AODV, DSR, and TORA. AODV and DSR have high end-to-end delay during the initial stages of simulation, but later on it decreases. AODV has the highest throughput. TORA values are linear. In the second scenario, the TORA routing protocol has high end-to-end delay and high media access delay, while AODV has the highest throughput.

Distributed Bellman Ford (DBF), DSR, and associativity based routing (ABR) are compared in [12]. Various parameters like control overhead, throughput, average end-to-end delay, and average hop-by-hop delay are considered. DBF routing protocol has highest overhead while DSR has lower routing overhead. DBF has fewer throughputs due to high usage of channels for route update messages. The throughput of ABR is higher than DSR due to the selection of paths that are associatively stable and that have a lighter load. The end-to-end delay of DBF is more or less the same as that of ABR and DSR but has high hop-by-hop delay at increasing mobile speeds.

A detailed analysis of DSDV routing protocols is done in [13]. The DSDV protocol is analyzed by varying the size of network, mobile density, pause time, and mobility speed. The overhead of the DSDV routing protocol increases with the increase in network size. By increasing the number of nodes, the number of packets lost gets decreased, thereby increasing the packet delivery. Increasing the node speed along with the number of connections results in an increase in the amount of packets delivered.

3.3 Properties Desired in a Routing Protocol

According to the RFC 2501 recommendation, the following characteristics are desired in a routing protocol [14]:

- Routing protocols should be loop free. Otherwise, there might be circulation of packets infinitely in the network. This may consume the bandwidth, which itself is available scarcely. This invalid circulation of packets may also lead to collisions with legitimate packets. This problem may be overcome by specifying the time to live (TTL) for each packet. Still, it's better to avoid this invalid circulation of packets altogether.
- Distributed operation is an inherent property for ad hoc networks. With distributed operation, communication can be established and carried out on the fly without any centralized communication mechanisms.
- A routing protocol should establish the routes only when it is required. The traffic distribution should be on demand. Thus, periodic broadcasting of packets gets avoided in an on-demand routing protocol, thereby reducing the control overhead in the network. On the flip side, it increases the delay in establishing the routes between source and destination node.

- When the delay introduced by an on-demand routing protocol is high, then proactive routing protocols can be considered. If the available bandwidth is able to sustain a proactive routing protocol, then in such situations proactive routing protocols are preferred.
- Both unidirectional and bidirectional links should be supported by the routing protocols. A majority of the routing protocols are assumed to work under bidirectional link while ignoring unidirectional links. But in real world, this seldom happens. A routing protocol should also be capable of exploiting the unidirectional links for improvement in performance.
- *Sleep mode* is permitted in ad hoc network nodes to conserve energy. If some of the nodes are not transmitting any data, then they are configured to go into sleep mode automatically and wake up when there is a need to broadcast the packets. A routing protocol should accommodate the sleep mode functionality of nodes. A routing protocol should wake up the nodes from their sleep mode when there are packets to be transmitted and should not consume too much energy for this functionality. Also, there should not be adverse effect on the overall performance of the network while waking up these sleeping nodes.
- In a wireless network like ad hoc networks, all the data is passed through air waves and can be readily exploited if there is no well-versed security in place. All the packets traveling between different nodes should be authenticated and encrypted so that they can be protected from prying eyes.

3.4 Discussion of Various Routing Protocols

Routing protocols can be classified based on their proactive and reactive nature. In proactive or table-driven routing protocols, periodic exchange of routing table information takes place between the nodes in the network. Routing information is extracted from these routing tables to establish a path between the source node and the destination node. In a reactive or on-demand routing protocol, routes are established by flooding the messages throughout the network. This is done only when the routes need to be established. There is another special class of routing protocols (i.e., routing protocols based on geographic location). In this type of protocol, the routes are established by taking advantage of the current position of the nodes in a geographical area. In this section we give a description of various routing protocols studied for this chapter (Figure 3.1).

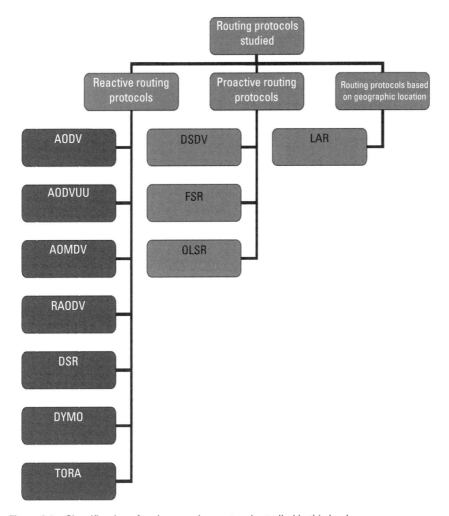

Figure 3.1 Classification of various routing protocols studied in this book.

3.4.1 Ad Hoc On Demand Distance Vector Routing Protocol

AODV routing protocol is a reactive routing protocol. A route is established using a route request/route reply query cycle. In order to set up a route to a destination for which it does not have a route, a source node broadcasts the route request (RREQ) packet across the networks. Nodes then update their information by receiving this RREQ packet and establish backwards pointers to the source node in the route tables. The RREQ contains the source nodes' IP address, current sequence number, and broadcast ID together with the most recent sequence number for the destination of which the source node is aware. By receiving the RREQ packet, a node may transmit a route reply (RREP) if it has either a route to the destination with the corresponding sequence number greater than or equal to that contained in the RREQ packet or it is the destination. A node receiving the RREQ packet forwards the packet only if it has not sent the packet previously. This unique RREQ ID helps in eliminating the duplicate RREQ packets. The RREQ packets create temporary route entries for the reverse path through every node they pass in the network. When they reach the destination, a RREP is sent back through the same path the RREQ was transmitted.

Figure 3.2 shows the establishment of the route from the source node 1 to the destination node 8. Routes between the source node and the destination node are maintained using HELLO messages, which are sent periodically. Whenever a route is used to forward the data packet the route expiry time is updated to the current time plus the active route timeout. An active neighbor node list is used by AODV at each node as a route entry to keep track of the neighboring nodes that are using the entry to route data packets. These nodes are notified with route error (RERR) packets when the link to the next hop node is broken. If a link is broken, then it is invalidated by the node that finds the broken route by sending the RERR to all its neighboring nodes. Every node maintains a route table entry that updates the route expiry time. A route is valid for the given expiry time, after which the route entry is deleted from the routing table [15].

3.4.2 Ad Hoc On Demand Distance Vector Routing Algorithm by Uppsala University (AODVUU)

AODVUU is the AODV routing protocol implementation by Uppsala University [16]. The main reason for using AODVUU protocol is to check whether AODVUU offers any performance improvement over the default AODV pro-

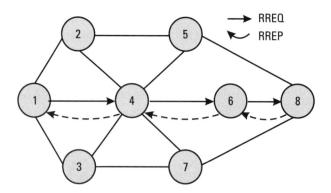

Figure 3.2 Route establishment in AODV routing protocol.

tocol available in the NS2 simulator. AODVUU is also an RFC 3561–compliant routing protocol.

3.4.3 Ad Hoc On Demand Multipath Distance Vector Routing Algorithm

The ad hoc on demand multipath distance vector routing algorithm is proposed in [17]. AOMDV employs the *multiple loop-free and link-disjoint path* technique. In AOMDV only disjoint nodes are considered in all the paths, thereby achieving path disjointness. For route discovery route request packets are propagated throughout the network, thereby establishing multiple paths at a destination node and at the intermediate nodes. Multiples loop-free paths are achieved using the advertised hop count method at each node. This advertised hop count is required to be maintained at each node in the route table entry. The route entry table at each node also contains a list of next hops along with the corresponding hop counts. Every node maintains an advertised hop count for the destination. The advertised hop count can be defined as the maximum hop count for all the paths. Route advertisements of the destination are sent using this hop count. An alternate path to the destination is accepted by a node if the hop count is less than the advertised hop count for the destination. We have used the AOMDV implementation for NS2 provided by [18].

3.4.4 Reverse Ad Hoc On Demand Distance Vector Routing Algorithm (RAODV)

In AODV route reply is used to construct a path from the source node to the destination node through various intermediate nodes. If any one of the intermediate nodes in the path moves out of the transmission range, then the path is broken. To overcome this advantage, RAODV is proposed in [19]. RAODV discovers many reverse routes from the source to the destination. In RAODV, the route discovery mechanism from the source node to the destination node is the same as that of the AODV routing protocol (i.e., a route request message is flooded throughout the network). Whenever an intermediate receives the route request message, then the message is forwarded to the next neighboring node. The intermediate nodes check whether they have received the same message based on the broadcast id and the sequence number. When the destination node receives the route request message, then it again floods the network with reverse route request message. The same procedure as mentioned previously during the route discovery from source node to the destination node is applied now by using reverse route request from destination node to the source node. If an intermediate node in the reverse path goes out of the transmission range when the route is discovered, the route error message is generated, which enables the source node and the destination node to choose alternate paths. When many paths are discovered from the destination node to the source node, then the best path is selected based on the sequence number and the least hop count from the destination node to the source node.

3.4.5 Dynamic Source Routing Algorithm (DSR)

DSR is an on-demand reactive routing protocol [20]. DSR is different from other routing protocols; it knows all intermediate nodes between source and destination, as it does not require sending of periodic HELLO messages for the maintenance of neighboring nodes. The route request (RREQ) message is broadcasted throughout the network by the source node. This RREQ message consists of unique RREQ ID and a list of all the intermediate nodes. A RREQ is forwarded if the node has not sent the RREQ message previously and if the address is not intended to itself. This duplicate RREQ messages are detected using the unique RREQ ID. When an intermediate node forwards the packets, it adds its address to the list. This address indicates the path the RREQ packet has traversed through various intermediate nodes to reach the destination node. When destination node 5 receives the RREQ packet, then the route from source node 1 to destination node 5 is established through different paths, as shown in Figure 3.3. Destination node 5 then generates a route reply (RREP) packet. This RREP packet then reaches the source node by making use of the addresses of the various intermediate nodes found in the list of the RREQ packet. When node 2 receives the RREP from destination node 5, it also contains the information about the various routes that are established between the source node and the destination node. The different paths between source node 1 and destination node 5 are 1-2-3-5 and 1-2-4-5. The source node then makes entry of all the available routes between the source node and the destination node in its route cache.

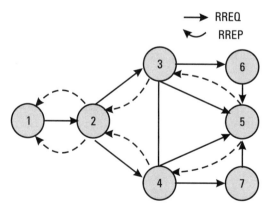

Figure 3.3 Route establishment in DSR routing protocol.

Whenever a packet is sent from a node to its neighboring node, the sender node decides whether or not the receiving node has received the packet successfully by applying the following mechanism: Consider Figure 3.4. Here, node 1 confirms that node 2 has received the packet when it hears 2 sending a packet to node 4. If a path between any two nodes in the route is broken then a route error (RERR) is generated, which is sent back to the source node. Various optimizations are incorporated for discovering the route and for its maintenance. If a route to the source node is present in the route cache of the destination node, then the destination node can make use of it to send the RREP packets back to the source node. Also, by piggybacking on the RREP messages, the destination node can establish a route to the source node. DSR can operate in promiscuous mode and then extract the information from the various packets that are generated in the network, and by using this information it can establish a route between the source node and the destination node.

3.4.6 Destination Sequence Distance Vector Routing Algorithm (DSDV)

The destination sequence distance vector routing algorithm proposed in [21] is based on the Bellman-Ford algorithm. Each of the nodes in the network maintains its own routing table. The packets are transmitted by making use of the information stored in the routing table. The routing table consists of the available destination nodes and the number of hops required in reaching the destination. Each of the routing tables is paired with a sequence number that is generated by the destination node. Any node that generates a sequence number will be an even number. If an odd sequence number is generated, then it indicates that it is ∞ metric. On finding new significant information, the routing information is broadcasted throughout the network by using packets. In DSDV a node advertises its routing table to all its neighboring nodes so that the available information is the latest and the node is able to locate all the other nodes in the

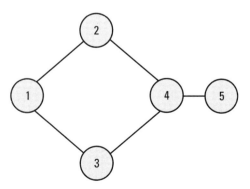

Figure 3.4 Packet confirmations for neighboring nodes.

network. The packets broadcasted by the source node consist of the addresses of the destination node, the number of hops needed to reach the destination node, and the sequence number generated by the destination node. A node checks its routing information available in the packet and compares the information with its routing table. If the destination address does not match with its own, then it increments the number of hops and passes the packet to the next available neighboring node. A broken link is indicated by the ∞ metric for the next hop destination. This information is updated throughout the network by broadcasting the packets. The routing information for a table can be updated either as a *full dump* or through *incremental* updates. The incremental update makes use of one network protocol data unit (NPDU), while full dump update requires multiple network protocol data units. When a node receives routing information, its sequence number is compared with the sequence number available in the table. If the sequence number available in the packet is greater than the sequence number found in the table, then the information is updated. Otherwise, the information is discarded.

3.4.7 Dynamic MANET On Demand (DYMO) Routing Protocol

DYMO routing protocol is an improved version of AODV routing protocol. Routes are established on demand. When a source node needs to establish a route to the destination, route request messages are flooded throughout the network. During broadcasting, only those nodes that have not broadcasted previously will forward the messages. The RREQ message includes its own address (source address), sequential number, a hop count, and the destination node address. A hop count of one is added to a RREQ packet originating at the source node. Each node that forwards the RREQ packet adds its own address and the sequence number. After the source node has sent the RREQ packet it waits for RREQ_WAIT_TIME duration. This is a constant value and is fixed at 1,000 seconds. In general simulations are done in 500s to 1,000s, and it is proven that these values are reasonably sufficient—if it is less than 500s, the network may not be stable, and if it exceeds 1000s, it is waste of time and does not influence the result. As the simulation time increases, it consumes lot of CPU resources and power.

Processing of RREQ by other nodes is done as follows: When a node receives the RREQ message, it is compared with its own routing table. If the routing table does not contain any information regarding the source node, then an entry is created for the source node. The next hop is to the node from which it had received the packets. This helps in establishing a route back to the originator.

Since the network is flooded with RREQ messages, almost all the nodes do contain information regarding the source node. If an entry is found to the source node, the node that had received the RREQ packet compares its sequence number and hop count with the information found in its own routing table. If an entry exists and if it is not stale, then the node updates its routing table with the latest information. Otherwise, the RREQ is dropped. Every node that evaluates a valid RREQ is capable of creating a reverse route to all the nodes whose addresses are found in the RREQ. After updating the route table, the node increments the hop count by a value of one to imply the number of hops the RREQ has traveled. Once the packet reaches the destination, a RREP message is sent back to the source node through the reverse path that had been created. This RREP packet contains the information regarding its sequence number, number of hop counts, and addresses. The same information is added by all the nodes that process the RREP along the reverse path.

Consider Figure 3.5. Whenever there is a break in the path, it has to be modified by generating a RERR message. Consider that the path between node 4 and node 6 is broken. Then node 4 has to generate RERR message containing the address and sequence number of the node that cannot be reached. This RERR is broadcasted throughout the network. When node 2 receives the RERR message, it compares the information found in the RERR message with its own routing table entries. If it has an entry, then the route information has to be removed if the next hop node (i.e., node 4) is same as the node from which it had received the RERR message. When node 3 and node 5 receive the RERR message, the information provided in RERR is checked with their corresponding entries. Since they do not use node 4 to reach 6, the RERR message is discarded. By broadcasting the RERR message, the concerned nodes are informed about any breakage in the path to the destination node [22–24].

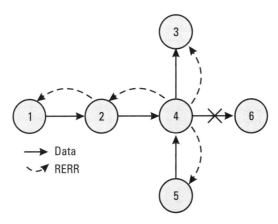

Figure 3.5 Route breakage scenario in DYMO [41].

3.4.8 Fisheye State Routing Protocol (FSR)

The FSR protocol employs a concept called *fish eye* technique. The idea behind fish eye technique is that a fish's eye is able to observe a maximum number of pixels near its focal point. The amount of pixels observed decreases as the distance increases from the focal point. If we think of the same concept from a network point of view, it means that a node maintains a high quality of information about its immediate neighboring nodes while the information decreases with the increase in the distance. A node maintains up-to-date information of the link table by obtaining the latest values from its neighboring nodes. This information is exchanged with the neighboring nodes periodically. These periodical exchanges are not event driven, thus reducing the overhead. When the nodes exchange the information, the entries with the highest sequence number replace the entries of a node with the lowest sequence number.

FSR routing protocol uses something called *scope*. The scope of a network can be defined as the group of nodes among whom communication takes place frequently, as they are within the assumed number of hops (or less number of hops). In Figure 3.6, for source node 1 all the neighboring nodes have one hop except from node 2. For source node 2, the number of hops is two for nodes 3, 4, and 5, while it is one for node 1 and 6. For source node 3, the number of hops from node 2, 4, and 5 is two, and three hops from node 6, while the number of hops from remaining nodes is less than one. For source node 4, the number of hops from 2, 3, and 5 is two. and for 6 it is three while from the remaining nodes it is less than two. For source node 5, the number of hops is

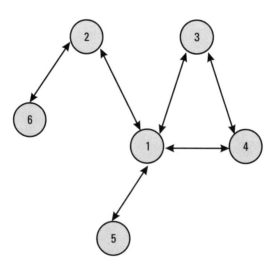

Figure 3.6 FSR demonstration [41].

less than two only for node 1. For source node 6 the number of hops is less than two only for node 1. These results are obtained by assuming that the scope is less than two. The nodes with less than two hops are the nodes with whom the packets are exchanged at a higher rate of frequency than the nodes having a hop count of two or more. This may result in keeping improper information regarding the nodes that are far away. But as the packet tends to get closer and closer to the destination node, the routes are established [25–27].

3.4.9 Location-Aided Routing (LAR) Protocol

Location-aided routing (LAR) makes use of the global positioning system. Using GPS, the LAR routing protocol is able to obtain information regarding the location of a node. In a WSN, the location of each node plays an important role after tracing or event detection on a particular node. At the time of deployment, manual configuration of a location about each node in wireless sensor environments is not feasible. But localization based on an algorithm is controlled by the directional antenna. In a sensor network, every node is configured with directional antenna, and few sensor nodes have GPS receivers. Such nodes having GPS receivers are called *anchor nodes*. These nodes help other normal nodes to localize them, based on the reference information provided by these anchor nodes. For localization, software is used that acts in two phases: The first one queries the sensor network and gathers information about neighbors, distance, and angle. The locations of nodes are determined with the help of this information when it is given as input to the software. The second one is used to compile the software for sensor nodes. The sensor nodes will calculate the location on installation of the software in sensor nodes. It is costly to have a GPS receiver on every sensor nodes and hence a limited number (based on requirement) of nodes have GPS receivers. The anchor nodes are assumed to know their own locations, whereas other normal nodes discover their location based on the reference information given by the anchor nodes. The anchor node broadcasts the reference position information several times with different power levels (so that normal nodes at different distances can receive the signal), and the normal sensor node calculates its position by measuring several parameters, like arrival direction and the strength of the signal, time of arrival and time difference of arrival, and angle of arrival and antenna directivity. However, the localization system should be power-aware, cheap, and variably accurate depending on the application.

In the LAR scheme, routes are established by using the flooding in an intelligent way. When a path is broken or when a node needs to establish a route to the destination node, then it initiates the route discovery by using route request messages.

When an intermediate node receives the same RREQ from two different nodes, then it discards one of the messages and forwards the other message. When an immediate node receives a RREQ message, it compares the address with its own. On finding that it is the same intended destination node, it triggers a route reply in the reverse direction. A timeout is specified within which the source node should receive the RREP packet. Otherwise, the source node will trigger the route discovery process once again. In this process the route RREQ is sent to every node. In LAR the number of packets sent to other nodes is reduced using various schemes.

LAR makes use of two concepts called *expected zone* and *request zone*. An expected zone can be defined as a circular region within which the destination node is expected to be present. The source node determines the expected zone of the destination node based on the average speed and the past location of the destination node. If the actual speed of the destination node is more than the average speed, then the destination node may be farther from the expected zone. Thus, the expected zone is just an estimation region in which the destination node may be present. If the source node is not able to obtain any information regarding the previous position of the destination node, then the entire network is considered the expected zone. Thus, the rule is that the more information is collected regarding the destination node, the more accurate the expected zone is.

The request zone is defined as the rectangular region within which a node forwards a packet. If a node is outside the request zone, then the node discards any packet it receives. Usually the request zone includes the expected zone. If the source node is not able to find a path to the destination node, then this request zone size is increased. The size of the request zone is dependent on the "average speed of the node and the time elapsed since the last position of the destination node was found out." In the LAR 1 scheme shown in Figure 3.7(a), the nodes 1, 3, 4, and 6 are present within the request zone and node 5 is the destination node present within the expected zone. When source node 1 sends a RREQ packet, it is broadcasted to the neighboring nodes 2, 3, and 4. Since node 2 is not present in the request zone, it discards all the received packets. Nodes 3 and 4 broadcast the packets to node 5 since they are found within the request zone. Once the destination node receives the RREQ, it triggers the RREP message containing information regarding its location. In LAR scheme 2, it is assumed that the location of the destination node is known to the source node. The source node sends a RREQ message that contains the location information of the destination node and the distance from the source node to the destination node. When an intermediate node receives this RREQ, it checks the distance mentioned in the RREQ message to its own value. If the distance is less than the mentioned distance value, then the RREQ message is forwarded or the RREQ message is discarded. In Figure 3.7(b), node 2 on receiving the

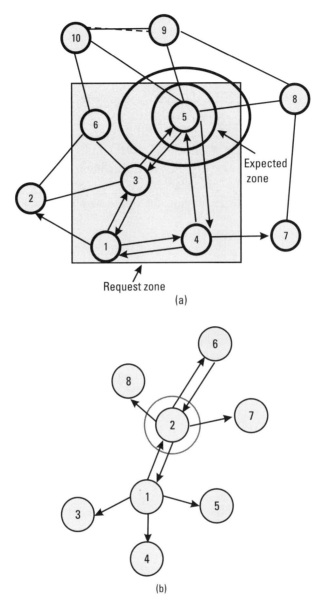

Figure 3.7 (a) LAR 1 scheme; (b) LAR 2 scheme [41].

RREQ compares its distance with its own routing table information. Since the distance is less, the RREQ packet is forwarded. When node 6 receives the RREQ packet, it finds that it is the destination node and it triggers a RREP in the reverse direction so a route is established to the source node [29–31].

3.4.10 Optimized Link State Routing Protocol

Optimized link state routing protocol (OLSR) is a table-driven, proactive based routing protocol. Multipoint relay (MPR) nodes are used to optimize the OLSR routing protocol. By MPRs the number of packets broadcasted in the network is minimized. A node selects a set of one-hop neighboring nodes to retransmit its packets. This subset of selected neighboring nodes is called the multipoint relays of that node. The MPR nodes are the only nodes those forward the packets during broadcasting. All the links between the nodes are assumed to be bidirectional. The MPR node is chosen in such a way that the chosen node is one hop and this one hop node also covers those neighboring nodes, which are two hops away from the originating node. The MPR nodes are affiliated to this original node. This reduces the number of messages that needs to be retransmitted.

In Figure 3.8, node 4 is two hops away from node 1. So node 3 is chosen as an MPR. Any node that is not present in the MPR list does not forward the packets. Every node in the network maintains information regarding the subset neighboring nodes that have been selected as MPR nodes. This subset information is called as MPR selector list.

Optimization in OLSR is achieved in two ways. First, the amount packets broadcasted in the network is reduced as only a selected few nodes called MPR broadcast the packets. Second, the size of the control packets is reduced as the

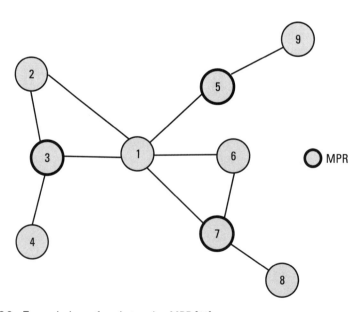

Figure 3.8 Transmissions of packets using MPR [41].

information regarding its multipoint relay selector set is provided instead of when providing an entire list of neighboring nodes [32–37].

3.4.11 Temporally Ordered Routing Algorithm (TORA)

TORA falls under a category of algorithms called *link reversal algorithms*. TORA is an on-demand routing protocol. Unlike other algorithms, the TORA routing protocol does not use the concept of shortest path for creating paths from source to destination, as it may itself take huge amount of bandwidth in the network. Instead of using the shortest path for computing the routes, the TORA algorithm maintains the direction of the next destination to forward the packets. Thus, a source node maintains one or more *downstream paths* to the destination node through multiple intermediate neighboring nodes. TORA reduces the control messages in the network by having the nodes query for a path only when they need to send a packet to a destination. In TORA, three steps are involved in establishing a network: (a) creating routes from source to destination, (b) maintaining the routes, and (c) erasing invalid routes. TORA uses the concept of directed acyclic graph (DAG) to establish downstream paths to the destination. This DAG is called as destination oriented DAG. A node marked as destination oriented DAG is the last node or the destination node, and no link originates from this node. It has the lowest height. Three different messages are used by TORA for establishing a path: the query (QRY) message for creating a route, update (UPD) message for creating and maintaining routes, and clear (CLR) message for erasing a route. Each of the nodes is associated with a height in the network. A link is established between the nodes based on the height. The establishment of the route from source to destination is based on the DAG mechanism, thus ensuring that all the routes are loop free. Packets move from the source node having the highest height to the destination node with the lowest height. It's the same top-down approach. When there is no directed link from source to destination, the source node triggers the QRY packet. The source node (node 1) broadcasts the QRY packet across all the nodes in the network. This QRY packet is forwarded by all the intermediate nodes that may contain a path to the destination.

Consider Figure 3.9(a): when the QRY packet reaches the destination node (node 9), then the destination node replies with a UPD message. Each node receiving this UPD message will set the value of the height to a value greater than the height of the node from which it had received. This results in the creation of the directed link from the source to the destination. This is the concept involved in the link reversal algorithm. This enables us to establish a number of multiple routes from the source to destination. Assume that the path between node 5 and node 6 is broken (Figure 3.9(b)). Then node 6 generates a UPD message with a new height value within a given defined time. Node 3

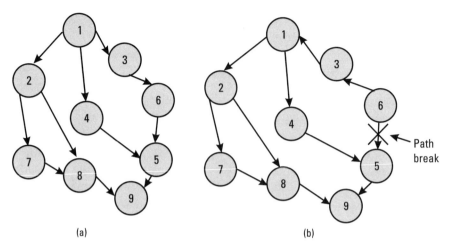

Figure 3.9 Directed paths in TORA [41].

reverses its link on receiving the UPD message. This reverse link indicates that the path to destination through that directed link is not available. If there is a break between node 1 and node 3, then it results in a partition of the network where the resulting invalid routes are erased using the CLR message [2, 38, 39].

3.5 Network Simulator, NS-2

NS-2 stands for network simulator version 2 and is the most popular discrete event tool used for the simulation of networking research. It provides substantial support for simulation of wired and wireless networks, satellite networks, transport control protocol, and routing such as TCP, UDP, FTP, HTTP, and DSR. Currently, NS-2 development is supported through DARPA with SAMAN and through NSF with CONSER, both in collaboration with other researchers, including ACIRI. This simulator was first developed for the simulation of general network. NS-2 simulator maintains list of events and executes one event after another. A single thread of control and no locking or race conditions are there. Simulations are based on the combination of C++ and OTcl. In general, C++ is used to provide the implementation of protocols and to extend the NS-2 library. OTcl is used to create and control the simulation environment itself, including the selection of output data. The simulation is run at the packet level. C++ is used for byte manipulation, packet processing, algorithm implementation, and where run-time speed is important, as the run time of C++ code is faster than that of OTcl.

The NS-2 provides a split-programming model. OTcl is interpreted and used to define the composition of the objects in the simulation (nodes, links, and so on) to allow changing scenarios without having to recompile. NS-2 sensor simulation is a modification of their mobile ad hoc simulation tools, with a small number of add-ons. Support is included for many of the things that make a sensor network unique, including limited hardware and power. An extension was developed in 2004 that allows for external phenomena to trigger events. NS-2 extensibility is popular for sensor networks. In addition to the various extensions to the simulation model, the object-oriented design of NS-2 allows for straightforward creation and use of new protocols. The NS-2 simulator produces a detailed trace file and an animation file for each network simulation, which is very convenient for analyzing the behavior. It is open source software downloadable from the Internet. For additional functionality, existing protocols can be extended, and new protocols can be implemented. Complex traffic patterns, topologies, and dynamic events can be automatically generated to test the created network topologies. The NS-2 can be run either in a UNIX-like operating system or Windows operating system with Cygwin [40].

3.6 Simulation Environment

Simulations are performed using NS-2. The simulated values of the radio network interface card are based on the 914-MHz Lucent Wave LAN direct sequence spread spectrum radio model. This model has a bit rate of 2 Mbps and a radio transmission range of 250m. The IEEE 802.11 distributed coordinated function with CSMA/CA is used as the underlying MAC protocol. Interface queue (IFQ) value of 70 is used to queue the routing and data packets.

The following metrics have been selected for evaluating the mobility models:

- Packet delivery ratio, defined as

$$\frac{\sum \text{Number of Received Data Packets}}{\sum \text{Number of Sent Data Packets}}$$

- Average network delay, defined as

$$\frac{\sum \left(\text{Time packet arrive at destination} - \text{Time packet sent at source} \right)}{\text{Total Number of Connection Pairs}}$$

- Throughput of the network, defined as

$$\frac{\sum \text{Node Throughputs of Data Transmission}}{\text{Total Number of Nodes}}$$

- Routing overhead, defined as

$$\frac{\sum \left[MAC\left(Control_{pkt}\right)_{sentsize} + MAC\left(Control_{pkt}\right)_{fwdSize} \right]}{\Delta T_{sim}}$$

- Average energy consumed, defined as

$$\frac{\sum PercentageEnergyConsumedbyallNodes}{NumberofNodes}$$

All the simulations scenarios are averaged for five different seeds while running independently. We have considered the effect of mobility speed, node density, and traffic load on the performance of AOMDV, DYMO, FSR, LAR, OLSR, and TORA routing protocols using community-based mobility model and SMS mobility model.

Both the mobility speed and the type of mobility model play a vital role in determining the performance of routing protocols. The mobility speed is varied from 5 m/s to 25 m/s in steps of 5. Here, these values are chosen because, for example, we can think of it as the speed of a moving vehicle in ad hoc networks. It is also the speed used in most of the literature and the speed from the best case to worst case. The routing protocols LAR, AOMDV, and TORA are compared using the community-based mobility model, and the routing protocols OLSR, DYMO, and FSR are compared using SMS mobility model. Scheme one implementation of route establishment is employed in LAR routing protocol. The effect of mobility speed and traffic load on the performance of OLSR and AOMDV routing protocols are studied using Levy walk mobility model and Gauss-Markov mobility model. The mobility speed of Levy model is varied from 1 to 3 m/s in steps of 0.5.

The number of nodes is varied from 10 to 50 in steps of 10. The goal is not to find the optimum number of nodes for the routing protocols but to see how these routing protocols fare against different node densities. Higher node density enables us to analyze whether these routing protocols are able to take the inherent advantages of a community model, especially the LAR routing protocol. The routing protocols LAR, AOMDV, and TORA are mapped against a number of nodes using the community based mobility model. At a higher node density, the connectivity between nodes is significantly better, resulting in better packet delivery. The parameters used for the simulation are depicted in Table 3.1.

3.7 Result Analysis

3.7.1 Evaluation of Performance Metrics

3.7.1.1 PDR with Mobility Speed

The LAR routing protocol has the highest packet delivery at varying mobile speed (Figure 3.10). This shows that a location based algorithm like the LAR routing protocol is able to take advantage of community model. The LAR

Table 3.1
Simulation Parameters

Simulator	NS2
Routing protocols	AOMDV, DYMO, FSR, LAR, OLSR and TORA
Mobility model	Community mobility model, SMS mobility model, Levy walk mobility model, and Gauss-Markov mobility model
Simulation time (sec)	900
Pause time (sec)	10
Simulation area (m)	1000 x 1000
Number of nodes	10, 20, 30, 40, 50
Transmission range	250 m
Maximum speed (m/s)	1, 1.5, 2, 2.5, 3 (Levy walk mobility model) 5, 10, 15, 20, 25 (other mobility models)
Traffic rate (pkts/sec)	5, 10, 15, 20, 25
Data payload (Bytes)	512

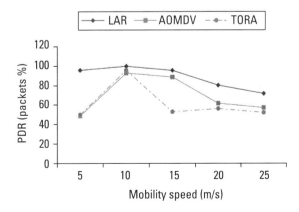

Figure 3.10 PDR vs. average mobility speed for LAR, AOMDV, and TORA routing protocols [41].

routing protocol maintains a delivery ratio of 98 percent to 70 percent. Only AOMDV can be comparable to the packet delivery capability of LAR. But there is an increases and decrease in the packet delivery of AOMDV, thereby indicating that it is not consistent with the mobility speed of the community model. The DYMO routing protocol has highest packet delivery. OLSR has an increase and decrease pattern but still is comparable with DYMO routing protocol. The packet delivery is around 90 percent (Figure 3.11). For the Levy walk mobility model, the packet delivery rates of AOMDV and OLSR are almost the same (Figure 3.12). The difference is insignificant. There is a slight increase and decrease pattern in the packet delivery. The OLSR routing protocol maintains the routing table information for all possible routes by periodically exchanging the HELLO and TC packets, thereby ensuring high packet delivery. For Gauss-Markov mobility model, the packet delivery ratio of AOMDV decreases with speed while for OLSR the packet delivery increases with speed. But for both, the routing protocols the delivery rate decreases after 20 m/s (Figure 3.13).

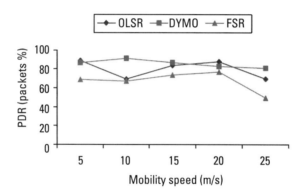

Figure 3.11 PDR vs. average mobility speed for OLSR, DYMO and FSR routing protocols [42].

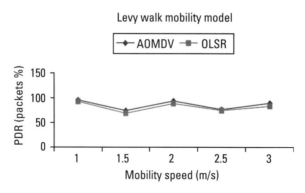

Figure 3.12 PDR vs. average mobility speed for AOMDV and OLSR routing protocols under Levy walk mobility model [42].

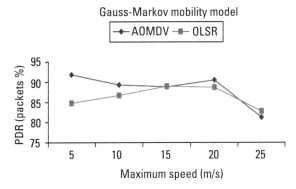

Figure 3.13 PDR vs. average mobility speed for AOMDV and OLSR routing protocols under Gauss-Markov mobility model [42].

3.7.1.2 Average Network Delay with Mobility Speed

The delay in the network increases with the increase in speed. This is in sharp contrast to what can be seen from the increase in node density. This shows that both node density and mobility speed play a vital role in determining the final outcome of the performance of a routing protocol (Figure 3.14). The delay of the TORA routing protocol is more than 0.1 sec, while the delay of LAR routing protocol is between 0.01 to 0.02 sec. The delay in TORA routing protocol may be mainly attributed to the creation of directed links in the network. The links are created across all the nodes in the network until the destination node is reached, resulting in higher delay in the network. The delay of the OLSR and DYMO routing protocols is between 0.01 to 0.02 sec. But the delay in FSR is very high. It increases with the increase in mobility speed. The delay of FSR is

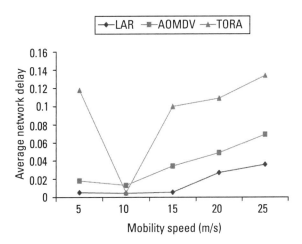

Figure 3.14 Average network delay vs. mobility speed for LAR, AOMDV, and TORA routing protocols [41].

in between 0.01 to 0.2 sec (Figure 3.15). The average network delay of AOMDV is higher than OLSR. The delay of AOMDV increases after 1.5 m/s from ~0.3 sec to ~0.05 sec. The average network delay is constant for OLSR and does not show much difference with the increase in speed. With the increase in speed, the broken paths may not be noticed by OLSR immediately. This may result in delay in the network (Figure 3.16). The OLSR routing protocol under Gauss-Markov mobility model does have the same delay as when applied under the Levy walk mobility model. The delay remains at ~0.2 sec as the speed increases (Figure 3.17).

3.7.1.3 Network Throughput with Mobility Speed

The network throughput of LAR routing protocol increases with increase in mobility speed (Figure 3.18). The results obtained for the LAR routing proto-

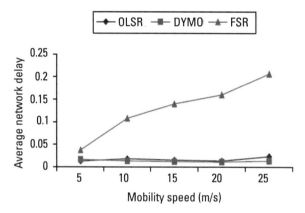

Figure 3.15 Average network delay vs. mobility speed for OLSR, DYMO, and FSR routing protocols [41].

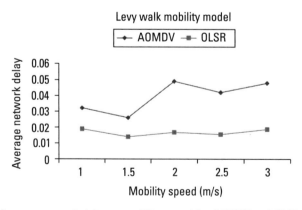

Figure 3.16 Average network delay vs. mobility speed for AOMDV and OLSR routing protocols under Levy walk mobility model [42].

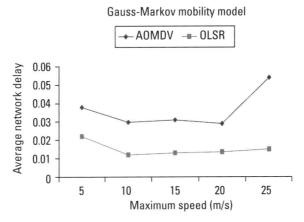

Figure 3.17 Average network delay vs. mobility speed for AOMDV and OLSR routing protocols under Gauss-Markov mobility model [42].

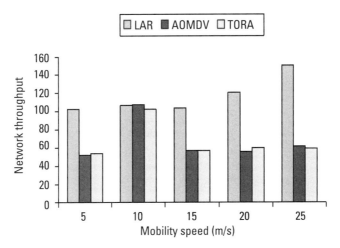

Figure 3.18 Network throughput vs. mobility speed for LAR, AOMDV, and TORA routing protocols [41].

col is ratified with [1], confirming our stand that the LAR routing protocol goes hand in hand with the community model. The throughput of AOMDV and TORA routing protocols remains from 40 to 50 Kbps, while the throughput of LAR routing protocol increases from 100 to 140 Kbps. The throughput of DYMO routing protocols remains steady, but it decreases in OLSR and FSR with increase in speed (Figure 3.19). The throughput results of the routing protocols are inconsistent with the packet delivery. In AOMDV the duplicate RREQ packets are not discarded altogether. Instead, each of the packets is checked for information to establish reverse paths from the destination node to the source node. Even though this may lead to the construction of multiple

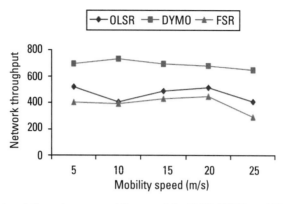

Figure 3.19 Network throughput vs. mobility speed for OLSR, DYMO, and FSR routing protocols [41].

paths, it also leads to delay, as each of the packets are analyzed while establishing multiple routes. Again as in packet delivery, the throughput of AOMDV and OLSR does not show much difference with the increase in speed (Figure 3.20). The throughput of AOMDV and OLSR increases with the increase in mobility speed. But the throughput decreases after 20 m/s (Figure 3.21). One surprising aspect that we uncovered is the amount of throughput achieved by these two routing protocols under Levy walk and Gauss-Markov mobility models. In the Levy walk model the simulation was conducted with less mobile speed as compared to Gauss-Markov mobility model. But there is remarkable difference in the amount of throughput achieved by these two routing protocols. The peak throughput for AOMDV and OLSR is around ~700 bps under Levy walk, and it is around ~350 bps and ~360 bps for AOMDV and OLSR under the Gauss-Markov mobility model.

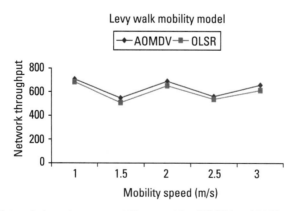

Figure 3.20 Network throughput vs. mobility speed for AOMDV and OLSR routing protocols under the Levy walk mobility model [42].

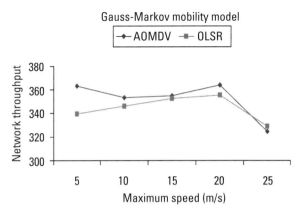

Figure 3.21 Network throughput vs. mobility speed for AOMDV and OLSR routing protocols udner Gauss-Markov mobility model [42].

3.7.1.4 Routing Overhead with Mobility Speed

The AOMDV and TORA routing protocols score over LAR routing protocol by having less overhead (Figure 3.22). The routing overhead in LAR is due to the fact that the expected zone is an estimated one, and sometimes the entire network is considered as an expected zone. This results in flooding the whole network with the route request messages. All the mechanisms employed in finding the destination zone optimally tend to fail in this case. This leads to clogging of the whole network, thereby increasing the routing packet overhead in the network. DYMO routing protocol is having higher overhead over OLSR and FSR, which was not expected considering that DYMO routing protocol performs better than AODV routing protocol (Figure 3.23). The routing overhead

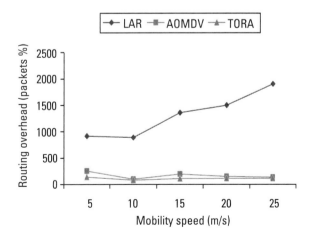

Figure 3.22 Routing overhead vs. mobility speed for LAR, AOMDV, and TORA routing protocols [41].

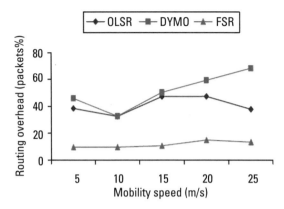

Figure 3.23 Routing overhead vs. mobility speed for OLSR, DYMO, and FSR routing protocols [41].

of DYMO ranges from 45 percent to 75 percent. This overhead in DYMO is mainly due to the flooding of route request to discover the destination node. This route request message is flooded throughout the network, thereby causing huge overhead. The overhead in OLSR is mainly due to the large selection of MPR resulting in large amounts of information being sent in the network. The routing overhead of AOMDV shows a zigzag pattern, while the routing overhead of OLSR is steady, indicating that the mobility speed of the Levy walk mobility model does not have much effect on the performance of OLSR (Figure 3.24). The overhead of OLSR is higher than AOMDV for Gauss-Markov, while AOMDV has higher overhead than OLSR under the Levy walk mobility model. An increase in mobility speed leads to an increase in broken routes.

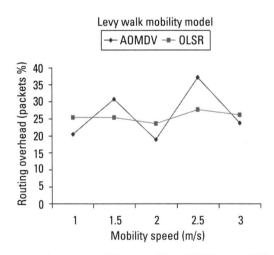

Figure 3.24 Routing overhead vs. mobility speed for AOMDV and OLSR routing protocols under Levy walk mobility model [42].

This results in the generation of HELLO messages for route establishment from source node to destination node, which dramatically increases the overhead in the network (Figure 3.25). OLSR has less overhead at lower mobility speeds.

3.7.1.5 Average Energy Consumed with Mobility Speed

The AOMDV and TORA routing protocols score over LAR routing protocol by having less energy consumption (Figure 3.26). Through simulation it has been shown in literature that TORA routing protocol consumes more energy than OLSR. But in the present situation with the community mobility model,

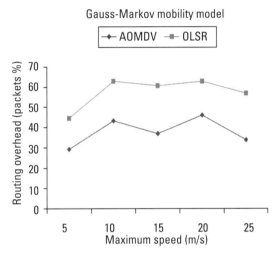

Figure 3.25 Routing overhead vs. mobility speed for AOMDV and OLSR routing protocols under Gauss-Markov mobility model [42].

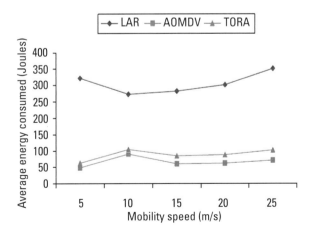

Figure 3.26 Average energy consumed vs. mobility speed for LAR, AOMDV, and TORA routing protocols [41].

the amount of energy consumed is almost the same for OLSR and TORA routing protocols. The energy consumption of the AOMDV and TORA routing protocols range of 50 to 100 joules, whereas the LAR routing protocol consumes energy at a rate of 300 to 350 joules. In the SMS mobility model, the energy consumption of all the routing protocols remains fairly the same with increase in mobility speed (Figure 3.27). The energy consumption of OLSR is higher than that of AOMDV, while AOMDV shows a steady behavior. The energy consumption of OLSR increases after 2m/s. The maximum energy consumption is ~200 joules (Figure 3.28). The energy consumption of AOMDV and OLSR show a straight-line behavior, consuming approximately the same amount of energy at varying mobile speeds (Figure 3.29).

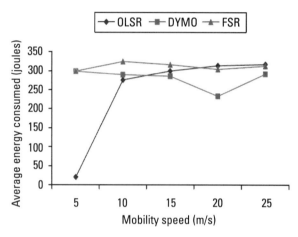

Figure 3.27 Average energy consumed vs. mobility speed for OLSR, DYMO, and FSR routing protocols [41].

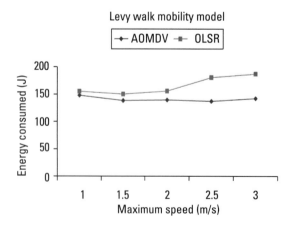

Figure 3.28 Energy consumed vs. mobility speed for AOMDV and OLSR routing protocols under the Levy walk mobility model [42].

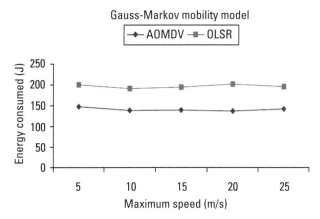

Figure 3.29 Energy consumed vs. mobility speed for AOMDV and OLSR rating protocols under the Gauss-Markov mobility model [42].

3.7.1.6 Evaluation of Performance Metrics with Traffic Load

When the number of nodes is less, the LAR routing protocol has a better packet delivery rate. But when the number of nodes is increased to more than 30, then all the routing protocols converge to the same level. At higher node density, all the routing protocols are able to provide nearly 100 percent packet delivery (Figure 3.30). The TORA routing protocol has the highest delay at a lower node density (Figure 3.31). But as the nodes increase, delay decreases among all the routing protocols. AOMDV has the least delay when compared to the LAR and TORA routing protocols. The delay in AOMDV is less due to the availability of alternate routes, thus avoiding the necessity of sending the route request packets again and again. It can be noticed that the throughput of all

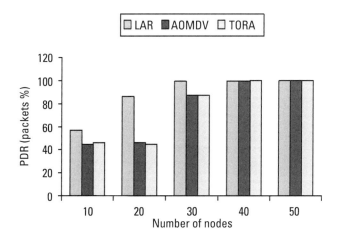

Figure 3.30 PDR vs. number of nodes for LAR, AOMDV, and TORA routing protocols [41].

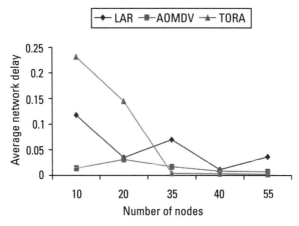

Figure 3.31 Average network delay vs. number of nodes for LAR, AOMDV, and TORA routing protocols [41].

the protocols increases with the increase in node density (Figure 3.32). This is due to increase in network connectivity. The routing overhead of the LAR routing protocol is astronomically high when compared to the AOMDV and TORA routing protocols. This can be attributed to the amount of route request packets that is flooded throughout the network. Even though the LAR routing protocol has various mechanisms to make the flooding of a network an intelligent one, the amount of packet needed to discover the destination node in the estimated zone itself still results in too much overhead in the network (Figure 3.33). The routing overhead of AOMDV is due to the additional route requests that are needed to maintain multiple paths in the network. The energy con-

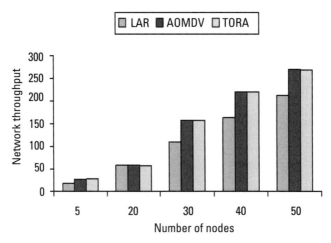

Figure 3.32 Network throughput vs. number of nodes for LAR, AOMDV, and TORA routing protocols [41].

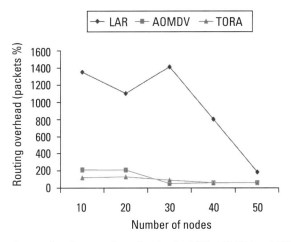

Figure 3.33 Routing overhead vs. number of nodes for LAR, AOMDV, and TORA routing protocols [41].

sumption of all the routing protocols increases with the increase in the number of nodes. More energy is consumed by the LAR routing protocol (Figure 3.34). The increase in the energy consumption in LAR routing protocol is due to the underlying flooding nature of the routing protocol. The simulation result sprang a few quite surprises, as we were expecting TORA to take advantage of its location capabilities when employed under the community mobility model. But here the best all-around performer was the AOMDV routing protocol. This shows that if were to employ a routing protocol for analyzing realistic mobil-

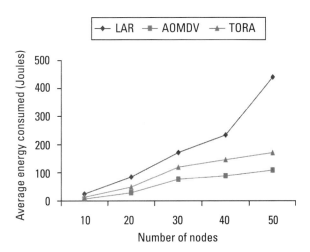

Figure 3.34 Average energy consumed vs. number of nodes for LAR, AOMDV, and TORA routing protocols [41].

ity models like community models, then AOMDV is the well-versed routing protocol.

3.7.1.7 PDR with Traffic Load

The traffic load is varied between 5 pkts/sec to 25 pkts/sec in steps of five. The OLSR, DYMO, and FSR routing protocols are analyzed using SMS mobility model with varying traffic loads. The packet delivery of OLSR and DYMO remains close to 100 percent up to 10 pkts/sec (Figure 3.35). With the increase in traffic load, the packet delivery of all the routing protocols tends to decrease. The DYMO routing protocol is able to cope with the congestion in the network better than OLSR and FSR routing protocols.

The packet delivery ratio for AOMDV and OLSR is approximately the same. There is a steep fall in packet delivery with the increase in traffic load. This shows that an increase in the number of packets causes congestion in the network, leading to the drop in the packets (Figure 3.36).

In the Gauss-Markov mobility model, the AOMDV routing protocol has the highest packet delivery. This is in contrast to Figure 3.40, where OLSR has the highest packet delivery. The packet delivery of both AOMDV and OLSR decreases with the increase in traffic load (Figure 3.37).

3.7.1.8 Average Network Delay with Traffic Load

The delay in FSR is the worst when compared to OLSR and DYMO. There is an astronomical increase in delay for the FSR routing protocol starting from 5 pkts/sec. For OLSR, the delay increases after 10 pkts/sec. The DYMO routing protocol has less delay when compared to other two routing protocols (Figure 3.38). The delay in the DYMO routing protocol increases after 15 pkts/sec. The delay varies from 0.1 sec to 0.4 sec. There is an astronomical increase in the delay after 10 pkts/sec for both the AOMDV and OLSR routing protocols.

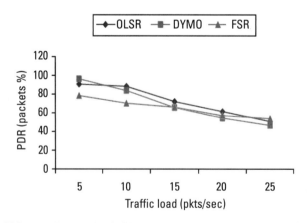

Figure 3.35 PDR vs. traffic load for OLSR, DYMO, and FSR routing protocols [41].

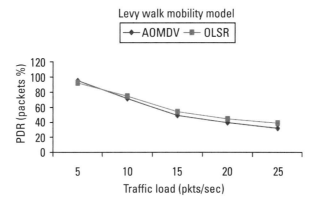

Figure 3.36 PDR vs. traffic load for AOMDV and OLSR routing protocols under the Levy walk model [42].

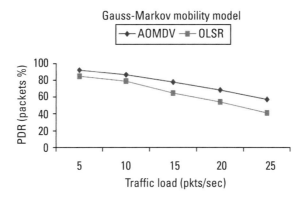

Figure 3.37 PDR vs. traffic load for AOMDV and OLSR routing protocols under Gauss-Markov mobility models [42].

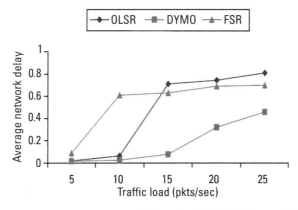

Figure 3.38 Average network delay vs. traffic load for OLSR, DYMO, and FSR routing protocols [41].

The delay in the AOMDV routing protocol is more than the OLSR routing protocol. There is an increase in the delay from 10 pkts/sec to 15 pkts/sec. After 15 pkts/sec, there is a stagnation in the delay. The peak delay for AOMDV is 1.2 sec, while for OLSR it's 0.6 sec (Figure 3.39). The difference in the network delay between AOMDV and OLSR is much less (Figure 3.40).

3.7.1.9 Network Throughput with Traffic Load

The DYMO has more packet delivery fails to maintain the momentum for network throughput (Figure 3.41). Surprisingly both the OLSR and FSR routing protocols score over DYMO in network throughput. The throughput of all the routing protocols increases with the increase in the traffic load. With the in-

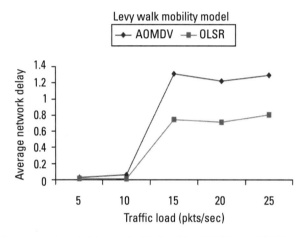

Figure 3.39 Average network delay vs. traffic load for AOMDV and OLSR routing protocols under the Levy walk mobility model [42].

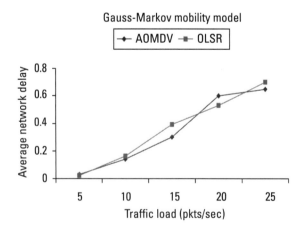

Figure 3.40 Average network delay vs. traffic load for AOMDV and OLSR routing protocols under the Gauss-Markov mobility model [42].

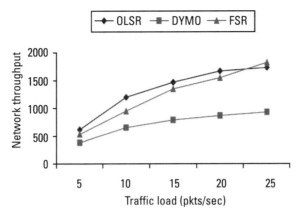

Figure 3.41 Network throughput vs. traffic load for OLSR, DYMO, and FSR routing protocols [41].

crease in traffic, there is an increase in throughput. This is quite expected due to the number of packets generated in the network. OLSR has a higher throughput than AOMDV for the Levy walk mobility model. The highest throughput achieved by AOMDV is around ~1000 bps, while it is ~1300 bps for the OLSR routing protocol (Figure 3.42). The throughput of AOMDV is higher than the OLSR routing protocol (Figure 3.43).

3.7.1.10 Routing Overhead with Traffic Load

The routing overhead of DYMO increases with an increase in traffic load, while the routing overhead of OLSR and FSR decreases with an increase in traffic load. FSR has less routing overhead. This factor can be attributed to the unique feature in FSR where it maintains the information of its immediate neighboring

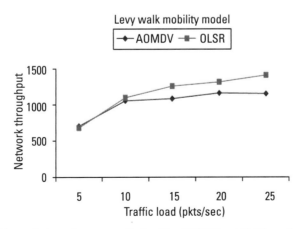

Figure 3.42 Network throughput vs. traffic load for AOMDV and OLSR routing protocols under the Levy walk mobility model [42].

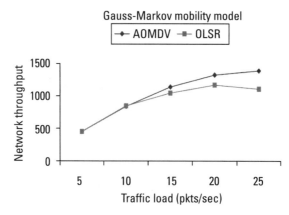

Figure 3.43 Network throughput vs. traffic load for AOMDV and OLSR routing protocols under the Gauss-Markov mobility model [42].

nodes and is thus able to find the destination node with a minimum number of hops (Figure 3.44). DYMO protocol's routing overhead increases up to 15 pkts/sec and maintains a steady state after that indicating that the network is saturated. The routing overhead for OLSR and FSR decreases with the increase in traffic load. The routing overhead for FSR decreases from 10 percent to 4 percent with the increase in traffic load. There is a decrease in the routing overhead of OLSR with the increase in the traffic load while the routing overhead of AOMDV is steady. AOMDV has more routing overhead than OLSR (Figure 3.45). The routing overhead of both AOMDV and OLSR decreases after the traffic load is increased to 10 pkts/sec (Figure 3.46). In AOMDV the intermediate nodes of various paths generate route reply packets. At higher traffic, the

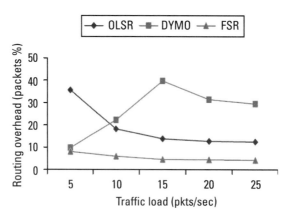

Figure 3.44 Routing overhead vs. traffic load for OLSR, DYMO, and FSR routing protocols [41].

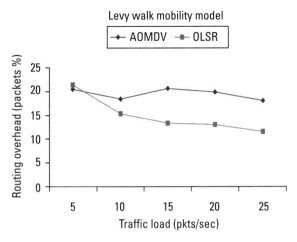

Figure 3.45 Routing overhead vs. traffic load for AOMDV and OLSR routing protocols under Levy walk mobility model [42].

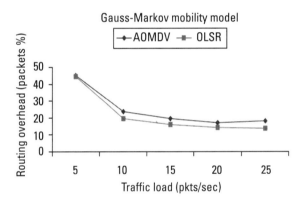

Figure 3.46 Routing overhead vs. traffic load for AOMDV and OLSR routing protocols under the Gauss-Markov mobility model [42].

source node increases the number of packets for route discovery, leading to an increase of overhead in the network.

3.7.1.11 Average Energy Consumed with Traffic Load

On observation, both OLSR and FSR have high energy consumption. Above 10 pkts/sec the energy consumption of OLSR increases at a very high rate. But the energy consumption of FSR routing is more than 300 joules starting from 5 pkts/sec (Figure 3.47). DYMO routing protocol maintains steady energy consumption between 150 to 200 joules. For the SMS mobility model, if high packet delivery, less delay, and less energy consumption are the criteria to employ a routing protocol, then the routing protocol of choice should be the

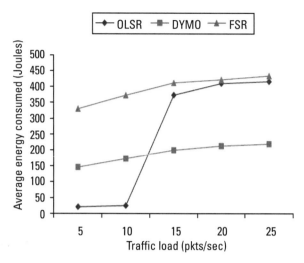

Figure 3.47 Average energy consumed vs. traffic load for OLSR, DYMO, and FSR routing protocols [41].

DYMO routing protocol. There is not much difference in the energy consumption of AOMDV and OLSR routing protocols with the increase in the traffic load (Figure 3.48). Again, there is an increase in the energy consumed with the increase in traffic load (Figure 3.49).

By observing the simulation results of performance metrics, it can be deduced that there is not much difference between AOMDV and OLSR regarding any of the parameters. This shows that the varying traffic load does not have significant impact on the performance of AOMDV and OLSR when deployed over the Gauss-Markov mobility model.

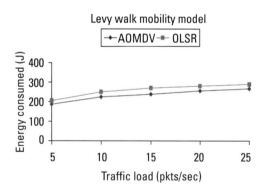

Figure 3.48 Energy consumed vs. traffic load for AOMDV and OLSR rating protocols for the Levy walk mobility model [42].

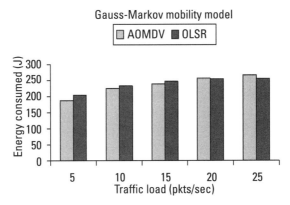

Figure 3.49 Energy consumed vs. traffic load for AOMDV and OLSR routing protocols for Gauss-Markov mobility model [40].

References

[1] Camp, T., and J. Boleng, et al., "Performance Comparison of Two Location Based Routing Protocols for Ad Hoc Networks." In *Proceedings of the IEEE 21st Annual Joint Conference of the IEEE Computer and Communications Societies (INFOCOM 2002)*, 2002, pp. 1678–1687.

[2] Prizada, A. A., et al., "Performance Comparison of Multipath AODV and DSR routing protocols in Hybrid Mesh Networks," In Proceedings of the IEEE International Conference on Networks (ICON06), Vol. 2, Singapore, September, 2006.

[3] Dwivedi, A. K., et al., "Performance of Routing Protocols for Mobile Ad Hoc and Wireless Sensor Networks: A Comparative Study," *International Journal of Recent Trends in Engineering*, Vol. 2, No. 4, November 2009.

[4] Cooper, N., and N. Meganathan, "Impact of Mobility Models on Multi Path Routing in Mobile Ad Hoc Networks," *International Journal of Computer Networks and Communications (IJCNC)*, Vol. 2, No. 1, January 2010.

[5] Nilsson, A., "Performance Analysis of Traffic Load and Node Density in Ad hoc Networks." In *Proceedings of the 5th European Wireless Conference (EW2004) Mobile and Wireless Systems*, Barcelona, Spain, February 24–27, 2004.

[6] Saad, M. I. M, et al., "Performance Analysis of Random Based Mobility Models in MANET Routing Protocol." *European Journal of Scientific Research*, Vol. 32, No. 4, pp. 444–454.

[7] Zhou, B., K. Xu, and M. Gerla, "Group and Swarm Mobility Models for Ad Hoc Network Scenarios Using Virtual Tracks," IEEE Military Communications Conference (MILCOM'04), Moneterey, CA, October 31–November 3, 2004.

[8] Jaap, S., M. Bechler, and L. Worf, "Evaluation of Routing Protocols for Vehicular Ad Hoc Networks in City Traffic Scenarios." In *Proceedings of the 5th International Conference on Intelligent Transportation Systems Telecommunications (ITST)*, Brest, France, June 2005.

[9] Broustis, I., et al., "A Comprehensive Comparison of Routing Protocols for Large Scale Wireless MANETs," 3rd International Workshop on Wireless Ad Hoc and Wireless Sensor Networks (in conjunction with SECON 2006), New York, June 28–30, 2006.

[10] Zhang, X., and G. R. Riley, "Performance of Routing Protocols in Very Large Scale Mobile Wireless Ad Hoc Networks," 13th IEEE International Symposium on Modeling, Analysis, and Simulation of Computer and Telecommunication Systems, Atlanta, GA, September 27–29, 2005.

[11] Chaudhry, S. R., and A. A. Khwildi, et al.,"A Performance Comparison of Multi On Demand Routing in Wireless Ad Hoc Networks." In *Proceedings of the IEEE International Conference on Wireless and Mobile Computing, Neworking and Communications (WiMob2005)*, Montreal, August 22–24, 2005.

[12] Lee, S.-J., C. K. Toh, and M. Gerla, "Performance Evaluation of Table Driven and On Demand Ad Hoc Routing Protocols." In *Proceedings of the IEEE International Conference on Personal, Indoor and Mobile Radio Communications (PIMRC1999)*, Osaka, Japan, September 12–15, 1999.

[13] Garousi, V., "Analysis of Network Traffic in Ad Hoc Networks Based on DSDV Protocol with Emphasis on Mobility and Communication Patterns." In *Proceedings of the IEEE International Conference in Central Asia on Internet (ICI 2005)*, Bishkek, Kygystan, September 2005.

[14] Corson, S., "Routing Protocol Performance Issues and Evaluation Considerations," Request For Comments: 2501, Network Working Group.

[15] Perkins, C., E. B. Royer and S. Das, "Ad Hoc On-Demand Distance Vector (AODV) Routing," RFC 3561, IETF Network Working Group, July 2003.

[16] AODVUU Implementation: http://core.it.uu.se/core/index.php/AODV-UU.

[17] Marina, M. K., and S. R. Das, "On-Demand Multipath Distance Vector Routing in Ad-hoc Networks." In *Proceedings of the IEEE International Conference on Network Protocols*, 2001, pp. 14–23.

[18] Caleffi, M., G. Ferraiuolo, and L. Pauro, "On Reliability of Dynamic Addressing Protocols in Mobile Ad hoc Networks." In *Proceedings of WRECOM'07, Wireless Rural and Emergency Communications Conference*, Rome, Italy, October 2007.

[19] Kim, C., E. Talipov, and B. Ahn, " A Reverse AODV routing protocol in Ad Hoc Mobile Networks," EUC Workshops 2006, LNCS 4097, 2006, pp. 522–531.

[20] Johnson, D. B., D. A. Maltz, and J. Broch, "DSR: The Dynamic Source Routing Protocol for Multi-Hop Wireless Ad Hoc Networks," Chapter 5, *Ad Hoc Networking*, Charles E. Perkins (ed.), Reading, MA: Addison-Wesley, 2001, pp. 139–172.

[21] Perkins, C. E., and P. Bhagwat, "Highly Dynamic Destination Sequenced Distance Vector Routing (DSDV) for Mobile Computers." In *Proceedings of International Conference on Communication Architecture, Protocols and Applications*, SIGCOm'94, London, England, pp. 234–244.

[22] Chakeres, I. D., and C. E. Perkins, "Dynamic Manet On Demand Routing Protocol," IETF Internet Draft, February 2008, draft-ietf-manet-dymo-12.txt.

[23] Thorup, R. E., "Implementing and Evaluating the DYMO Routing Protocol," Master's Thesis, Department of Comptuer Science, University of Aarhus, Denmark.

[24] Rahayu, S., and A. Aziz, et al., "Performance Evaluation of AODV, DSR and DYMO Routing Protocols in MANET," International Conference on Scientific and Social Science Research, Kaula Lampur, Malaysia, March 14–15, 2009.

[25] Pei, G., M. Gerla and T. W. Chen,"Fisheye State Routing: A Routing Scheme for Ad Hoc Wireless Networks." In *Proceedings of IEEE International Conference on Communications (ICC2000)*, New Orleans, LA, June 18–22, 2000.

[26] Royer, E. M., S.-J. Lee and C. E. Perkins, "The Effects of MAC Protocols on Ad hoc Network Communication." In *Proceedings of IEEE WCNC 2000*, Chicago, IL, September 2000.

[27] Sun, A. C. "Design and Implementation of Fisheye Routing Protocol for Mobile Ad Hoc Networks," http://www.scanner-group.mit.edu/PDFS/SunA.pdf.

[28] Murthy, C. S. R., and B. S. Manoj, *Ad Hoc Wireless Networks: Architectures and Protocols*, Second Edition, Englewood Cliffs, NJ: Prentice Hall.

[29] Ko, Y. B., and N. H. Vaidya, "Location Aided Routing (LAR) in Mobile Ad Hoc Networks." In *Proceedings of the 4th Annual ACM/IEEE International Conference on Mobile Computing and Networking*, Dallas, TX, 1998.

[30] Broustis, I., and G. Jakllari, et al., "A Comprehensive Comparison of Routing Protocols for Large Scale Wireless MANETs." In *Proceedings of the 3rd International Workshop on Wireless Ad Hoc and Sensor Networks*, New York, June 28–30, 2006.

[31] Sarkar, S. K., T. G. Basavaraju, et al, *Ad Hoc Mobile Wireless Networks, Principles, Protocols and Applications*, Auerbach Publications, 2008.

[32] Clausen, T., and P. Jacquet, "Optimized Link State Routing Protocol for Ad Hoc Networks," http://hipercom.inria.fr/olsr/rfc3626.txt.

[33] Frikha, M., and M. Maamer, "Implementation and Simulation of OLSR Protocol with QOS in Ad Hoc Networks." In *Proceedings of the Second International Symposium on Communications, Control and Signal Processing (ISCCSP'06)*, Marrakech, Morocco, March 13–15.

[34] Novatnack, J., and H. Arora, "Evaluating Ad Hoc Routing Protocols with Respect to Quality of Service," IEEE International Conference on Wireless and Mobile Computing, Networking and Communications (WiMob'05), October 3, 2005.

[35] Holter, K. "Comparing AODV and OLSR," folk.uio.no/kenneho/studies/essay/essay.html, last accessed 2010.

[36] Ge, Y., et al., "Quality of Service Routing in Ad Hoc Networks Using OLSR." In *Proceedings of the 36th IEEE International conference on System Sciences (HICSS'03)*, 2002.

[37] Park, V. D., and M. S. Corson, "A Performance Comparison of the Temporally Ordered Routing Algorithm and Ideal Link State Routing." In *Proceedings of the 3rd IEEE Symposium on Computers and Communications*, Washington, DC, 1998.

[38] Park, V. D., J. P. Macket, and M. S. Corson, "Applicability of the Temporally Ordered Routing Algorithm for Use in Mobile Tactical Networks," IEEE Military Communications Conference (MILCOM 98), 1998.

[39] Dutkiewicz, E., et al., "A Review of Routing Protocols for Mobile Ad Hoc Networks," *Ad Hoc Networks,* Vol. 2, Issue 1, January, 2004, pp. 1–22.

[40] Information Sciences Institute, "The Network Simulator Ns-2," http://www.isi.edu/nanam/ns/, University of Southern California, last accessed 2010.

[41] Proceedings of the International Journal of Computer and Network Security (IJCNS), Vol. 2, No. 5, May 2010, pp. 30–40.

[42] Proceedings of the Interantial Journal on Computer Science and Enegineering, Vol. 2, No. 4, July 2010, pp. 979–986.

4

An Empirical Study of Various Mobility Models in MANETs

4.1 Introduction

A characteristic feature of ad hoc networks is the infrastructureless and seamless connectivity of the wireless mobile nodes. Mobility plays an important role in the connectivity of these nodes. In this chapter, we study the performance comparison of various mobility models like community model, group force mobility model (GFMM), reference point group mobility (RPGM) model, Manhattan mobility model, random waypoint-steady state (RWP-SS) mobility model, random waypoint (RWP), random walk with reflection (RW-R), and random walk with wrapping (RW-W). Community models GFMM and RPGM are pure group mobility models, while Manhattan mobility model can be considered a pseudo group mobility model.

In a mobile ad hoc network, sometimes the mobile nodes move in groups instead of moving individually. For example, in the case of search and rescue operations in natural disasters it is necessary for the rescuers to work cooperatively and move in groups to accomplish a particular goal. Also, in battlefield scenarios a group of soldiers equipped with mobile devices may form a mobile ad hoc network. In many situations these soldiers need to work cooperatively and need to move in groups. In order to model such group-based movement and communication pattern, a group mobility model is needed. In such models a spatial dependency exists between the nodes that tend to move in a correlated manner. The difference between a pure group mobility and pseudo group mobility model is that in pure group mobility models each and every node is in some way attached to some group, but in pseudo group-based mobility models there can be many nodes that are not assigned to any group. Also the number of groups considered is less in pseudo group based mobility models when compared to pure group based mobility models.

The group community mobility model is based on the social network scenario. This model takes the social network as input and emphasizes the relation between any two members of a group or any individuals of the group. The group community model also conceptualizes the interaction between groups of a social community network. The mobile nodes with strong social relationships are more likely to be spatially co-located and often move within the same geographical area. This mobility model can be used to model humans moving in groups or groups of humans that are clustered together.

Random models are purely randomized (i.e., there is no mobility coherence between points in time n and $n+1$ is existent). Due to the highly dynamic nature of the mobile ad hoc networks the network topology changes frequently. The nodes move in highly unpredictable manner with randomly chosen speed and direction. In order to model their highly dynamic movement patterns, random mobility models are used.

We have included RWP-SS to give a whole picture of how group mobility models stand against a random model like RWP-SS. From our analysis we deduce that group mobility models hold inherent advantage over mobility models like random waypoint models. Among the group mobility models, the community model has good performance when compared to other mobility models. Energy consumption of these mobility models has also been analyzed. Various metrics like packet delivery ratio, average network delay, network throughput, routing overhead, and number of hops have been considered. The results obtained in this chapter colligates with the theoretical results in [1].

Mobility is the central theme in the connectivity of wireless devices. Mobile ad hoc networks provide the principality behind the ubiquitous computing. Mobile ad hoc networks do not have any centralized administration. All the nodes in an ad hoc network are autonomously connected in a dynamic manner.

In ad hoc networks, the topology changes very frequently due to the mobility of the nodes. Mobility plays a very important role in the performance of the routing protocol, and hence the underlying mobility model should be carefully selected for optimum performance.

This chapter presents the performance study of various group mobility models like community model, GFMM, RPGM, a pseudo group mobility model like Manhattan model, and a random mobility model like rwp-ss. In addition, this chapter provides comparison of mobility models like random waypoint, random walk with reflection, and random walk with wrapping. To the best of our knowledge, no work has been reported that compares and studies the performance of all these mobility models together.

4.2 Literature Background

The authors of [2] have considered various synthetic entity mobility models and group mobility models. Under synthetic entity mobility models, they have given a detailed description of several different mobility models: random walk mobility model, random waypoint mobility model, random direction mobility model, a boundless simulation area mobility model, and Gauss-Markov mobility model. Five group mobility models discussed are exponential correlated random mobility model, column mobility model, nomadic mobility model, pursue mobility model, and reference point group mobility model. The authors have also discussed the importance of selecting the underlying mobility model. The author simulated the results of four different mobility models and concludes that the random way point mobility model has the highest PDR and lowest end-to-end delay.

Fan Bai and Ahmed Helmy have conducted another survey of the various mobility models in ad hoc networks [3]. The authors categorize the various mobility models into random models, mobility models with temporal dependency, mobility models with spatial dependency, and mobility models with geographic restrictions, and they have discussed various mobility models under each category.

In [4], the authors have considered three mobility models: pursue mobility model, column mobility model, and RPGM-RW mobility models, and they have studied the effects of these mobility models on the performance of three ad hoc routing protocols, AODV, DSDV, and DSR. The authors conclude that the DSR protocol has the highest PDR under the pursue and random waypoint mobility model and AODV has lower routing overhead under column and RPGM-RW mobility models.

The authors of [5] have mapped AODV, DSR, and TORA routing protocols against random waypoint, random walk, and pursue mobility models. These mobility models are simulated using the OPNET simulation tool. The authors have compared these routing protocols under various mobility models by varying the mobility speed and the pause time.

A comparison of random waypoint and Gauss-Markov mobility models is done in [6], an analysis of DSR and DSDV routing protocols under various mobility models like RPGM, Manhattan mobility, and freeway mobility is done in [7]. The authors considered throughput the main metrics and mapped it against number of nodes and mobility speed.

4.3 Description of Various Mobility Models

Mobility models can be divided into trace-driven mobility models and synthetic mobility models. In trace-driven mobility models the node movement are based on realistic user movements that occur in the real world. But extracting the movements from real-world traces is highly difficult. Thus, traces have to be obtained over longer periods of time, which requires a significant amount of investment. In situations like this the synthetic mobility models are considered. The synthetic mobility models can be modeled on various scientific calculations, allowing researchers to evaluate the routing protocols under unintended scenarios that may be too difficult to extract from real traces. The synthetic mobility models are further classified into entity based mobility models and group based mobility models. In entity based mobility models, the movement of nodes is defined independently from one another, and they are not related to each other. In group based mobility models all the nodes move in a cluster of varying groups. In pure group based mobility models all the nodes are in some way attached to some group, but in pseudo group based mobility models there can be many nodes that are not assigned to any group. Also the number of groups considered is less in pseudo group based mobility model when compared to pure group based mobility model [8–10]. The classification of various mobility models studied for this chapter is given in Figure 4.1.

4.3.1 Random Waypoint Mobility Model

Random waypoint mobility model is one of the most popular mobility models. It was used as the underlying mobility model for almost all the early research work in mobile ad hoc networks. RWP comes under the category of synthetic mobility models. Paths in a RWP are straight line represented by $p(u) = (1-u)m_0 + um_1$, where m_0 and m_1 are the end points. RWP consists of two phases, namely, move and pause phases. Each of the mobile nodes is assigned with a destination and a speed. The mobile reaches the destination with the specified speed in a straight line. On reaching the destination, it pauses for a specified period and then again a new destination is determined [11–14].

4.3.2 Random Waypoint Mobility Model with Steady State

While considering the random waypoint mobility model for simulation, a dissimilar mobility pattern is observed during the initial mobility duration and at the later stage of the simulation. In literature, to avoid the mentioned situation, many of the studies follow a procedure where the initial few seconds are discarded and then it is assumed that the remaining seconds of the simulation have

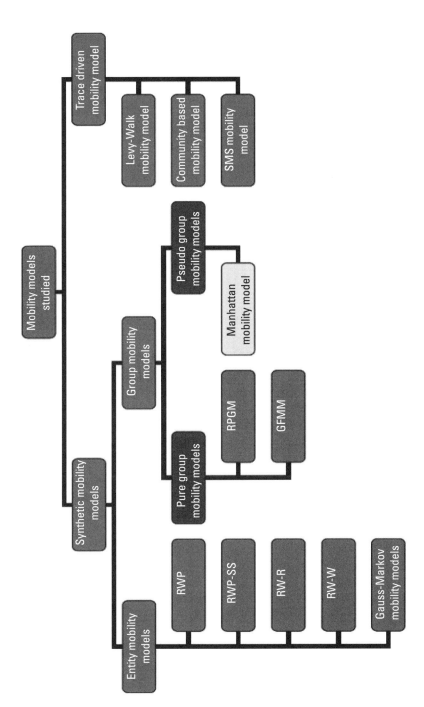

Figure 4.1 Classification of various mobility models studied.

a similar pattern. But this method is crude, as it cannot be told at which point the dissimilar pattern starts or stops. To overcome this problem, the authors of [15] have proposed, the random way point-steady state mobility model. The initial speed and the stationary distribution location are sampled to overcome the problem of discarding the initial simulation data. The RWP-SS without pause is given by:

$$F^{-1}(u) = \frac{S_1^u}{S_0^{u-1}} \qquad (4.1)$$

Here S is the initial speed chosen uniformly over (0, 1), and $F^{-1}(u)$ is the inverse of the cumulative distribution function. RWP-SS with pause is given by

$$H_0(p) = \frac{\int_0^p [1 - H(t)] dt}{E(p)} \qquad (4.2)$$

where, $H(p)$ is the cumulative distribution function, $E(p)$ is the expected length of a pause. (Equations 4.1 and 4.2 are from [15].)

4.3.3 Random Walk Mobility Model with Reflection

In the random walk mobility model with reflection, a node on touching the boundary (transmission range of the node, which is able to communicate) gets reflected back to its domain (a set of nodes that may be combined routers and hosts themselves forms the network routing infrastructure) [16]. The random walk mobility model reflection is given by the billiard reflection function

$$b : R^2 \to A$$

which is defined as

$$\begin{pmatrix} x \\ y \end{pmatrix} \to b \begin{pmatrix} x \\ y \end{pmatrix} = \begin{pmatrix} a_1 b_1 \left(\frac{x}{a_1} \right) \\ a_2 b_1 \left(\frac{y}{a_2} \right) \end{pmatrix}$$

where $b_1 : R \to [0,1]$ is the 2-periodic function: $b_1(x) = |x|$, for $-1 \leq x \leq 1$.

4.3.4 Random Walk Mobility Model with Wrapping

The special case of wrapping with random waypoint mobility model is discussed in [16]. To consider that a mobility model is having a perfect simulation model, it should not have a transient time. A trip combines the duration at any given point of time T_n with a P_n position will have a trip duration of $S_n \in R_+$. The next transition time for a mobile node is given by $T_{n+1} = T_n + S_n$ with a position of $P_n\left(\dfrac{t-T_n}{S_n}\right)$.

Let D, a closed, bounded, and connected subset of R^2, be the rectangle $[0, b_1] \times [0, b_2]$ where $R^2 \to D$ is the wrapping function (ω). A speed vector $\vec{V}_n$ and a trip duration S_n are chosen independently by the trip selection rule. Selection of the speed vector is the same as the selection of movement direction and a numerical speed. The node starts from the end point E_n in the direction at the rate provided by the speed vector. It is to be noted that wrapping does not change the speed vector. On reaching the end of a boundary (x_0, b_2) the node wraps to the other end, at a location of $(x_0, 0)$. The wrapping function is then expressed as:

$$\omega(x, y) = (x \bmod b_1, y \bmod b_2)$$

If not a pause the path M_n is represented by $(E_n, \vec{V}_n, S_n)$, where $M_n(f) = E_n + fS_n\vec{V}_n$) and where 'f' is a fraction such that $M(f)$ is the point on M attained when 'f' $\in [0,1]$ of the path is traversed.

4.3.5 Gauss-Markov Mobility Model

The Gauss-Markov mobility model [17, 18] makes use of the probability density function. The Gauss-Markov mobility model determines the future location of a node based on its velocity and the previous location of the node. The location of a node is made known to the entire network at any given point of time. When a node moves, it checks the distance it has covered. If the node has covered a minimum acceptable distance or more, then the node updates the network about its current position. To reach a node, the network inspects the previous location of the node. Based on the velocity and the previous location it sends a message in shorter distance order to locate the current position.

The velocity of a node is given by

$$V_n = \alpha^n v_0 + (1-\alpha^n)\mu + \sqrt{1-\alpha^2 \sum_{i=0}^{n-1}\alpha^{n-i-1}x_i}$$

$Vn = V(n\Delta t)$ and $\alpha = e^{-b\Delta t}$, where Δt = clock tick period (normalized to 1), v_0 = initial velocity, m = mean.

The Gauss-Markov mobility model can also be used to mimic other mobility models like fluid flow mobility model and random walk mobility model.

4.3.6 Reference Point Group Mobility Model

The reference point group mobility model is a popular group mobility model [19]. In this model, mobile hosts are organized by groups according to their logical relationships. There is random selection of a leader for the group in the RPGM mobility model. The authors call this step as selecting logical center (group leader). This group leader is used to set the speed, position, and direction of the group. The nodes in a group are usually randomly distributed around the reference point. All the nodes in the group follow this leader even though they have their own individual random motion behavior. The group leader selects a random destination or *checkpoint* and moves toward that checkpoint or destination at a given speed. New destination is selected using the motion path given to each group. This motion path is calculated using the checkpoints [20]. This general description of group mobility can be used to create a variety of models for different kinds of mobility applications such as military battlefield communications. For example, a number of soldiers may move together in a group or during disaster relief where various rescue members like firefighters, police officers, and medical assistants form different groups and work cooperatively.

4.3.7 Group Force Mobility Model

The group force mobility model proposed in [21] is based on the concept of attraction and repulsion of mobile nodes. The GFMM can be easily applied to human groups. The authors of GFMM compare these mobility models with RPGM by taking into consideration the various metrics proposed in the IMPORTANT framework. There is repulsion among human nodes to avoid collision among themselves and to other obstacles in their paths, while attraction is used to reach the destination. The GFMM introduces novel concepts called *loose group* and *tight group* (Figure 4.2). A group is loose if the distance between the hosts range from 0 to 15m or greater than 15m, while a group can be considered tight if the distance is in between 0 to 5m. The repulsive force or the exponential force denoted by Exppdf(x, μ) decreases as the nodes move apart farther. The attractive force is represented by three different models and the exponential force is subtracted from these three different attractive force models. The three different attractive force models are as follows (equations 4.3, 4.4, and 4.5 are from [21]):

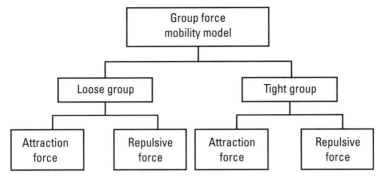

Figure 4.2 Steps involved in the establishment of the GFMM [21].

Chi Squared group force model:

$$\overline{f}_{ij} = A_{ij} \left[\frac{\overline{r}_i - \overline{r}_j}{d_{ij}} \right] x \\ \left[\text{Exppdf}(x,\mu) - \text{chi2pdf}(x,\mu) \right] \quad (4.3)$$

Rayleigh Group force model:

$$\overline{f}_{ij} = A_{ij} \left[\frac{\overline{r}_i - \overline{r}_j}{d_{ij}} \right] x \\ \left[\text{Exppdf}(x,\mu) - \text{Rayleighpdf}(x,B) \right] \quad (4.4)$$

Fisher-Snedecor or F group force model:

$$\overline{f}_{ij} = A_{ij} \left[\frac{\overline{r}_i - \overline{r}_j}{d_{ij}} \right] x \\ \left[\text{Exppdf}(x,\mu) - \text{Fpdf}(x,v_1,v_2) \right] \quad (4.5)$$

where:

A_{ij} and B are the simulation specific constants.
v_1, v_2 are positive integers.
X exists in the interval $[0 \ \infty]$

4.3.8 Manhattan Mobility Model

Manhattan is well planned and the most densely populated of the five boroughs of New York City. The Manhattan model is used to emulate the movement patterns of mobile nodes on streets defined by maps. It can be useful in modeling movement in an urban area where a pervasive computing service between portable devices is provided. The map is composed of a number of horizontal and vertical streets. The mobile node is allowed to move along the grid of horizontal and vertical streets on the map. At an intersection of a horizontal and a vertical street, a mobile node can turn left, right, or go straight with certain probability. There are internode and intranode relationships involved in the Manhattan model. Thus, the Manhattan mobility model is expected to have high spatial dependence and high temporal dependence. Some parameters generally used at each intersection are given here:

- Probability of moving on the same street is 0.5;
- Probability of turning right is 0.25;
- Probability of turning left is 0.25.

The mobile node speed is dependent on the direction of the previous movement [20–22].

4.3.9 Levy-Walk Mobility Model

The Levy-walk mobility model [23, 24] more or less imitates the human mobility traits in an outdoor condition. Real-world human mobility traces are generated at various places that include two university campuses, a metropolitan area, and a theme park by using GPS devices.

A flight is defined as the movement of an object along a straight line without any change in the direction. Various features like flight length distribution, pause time distribution, mean squared displacement, and velocity are analyzed for the real-world human mobility traces. To obtain a human walk flight from the traces is difficult, as the human seldom walks in a straight line. Also, there might not be continuity in a human walk, as he may pause for few minutes, or he may change the direction, or may move in a vehicle and disappear for few minutes and appear in another location, or there might be no battery in the GPS device. To reduce errors due to these factors, three different methods are proposed for analysis. They are rectangular, angle, and pause based methods.

The distance between any two points is considered as a flight in the rectangular model if there is no pause while moving between the two points and if the length between any two points is a perpendicular length to the point from that position. The angle model takes various flights found out from the rectangular model and combines them in to a single flight, provided that there is no pause between any of the successive flights and the relative angle is less than {insert in-line equation here} between any two consecutive flights. {insert in-line equation here} is a model defined parameter. The pause model also combines the flights obtained from the rectangular trajectory. It establishes more trajectories and accordingly represents the more natural human walk.

The Levy-walk model consists of four variables: flight length (l), direction (θ), flight time (Δt_f), and pause time (Δt_p). The Levy distribution with α and β coefficients is represented as follows:

$$f_x(x) = \frac{1}{2\Pi} \int_{-w}^{+w} e^{-itx - |et|^{\alpha}} dt \tag{4.6}$$

For $\alpha = 1$ it is Cauchy distribution, and for $\alpha = 2$ it is Gaussian distribution.

4.3.10 Community Based Mobility Model

A new social network based model called community mobility model proposed by [25] can be used to model humans moving in groups or groups of humans that are clustered together. The authors of the community model have evaluated their model by using real-time synthetic mobility traces provided by the Inter Research Laboratory, Cambridge. The community model can be conceptualized as shown in Figure 4.3.

The first step in establishing a community model is using the *social network as input* to the community mobility model. It involves two ways: *modeling of social relationships* and *detection of community structures*. Modeling of social relationship can be represented as a weighted graph matrix. If any of the elements in the matrix is greater than the specified threshold value, then that element in the graph is set to 1. If it is less than the threshold value, then it is set to 0. A value of 1 represents strong social interaction between the groups, and a value of 0 represents no interaction. The concept of 1 and 0 is used to emphasize the relation between any two members of a group or any individuals of the group. The next step is to conceptualize the interaction between groups of a social community network. The authors of the community model have implemented this aspect by considering an algorithm provided by [26]. Groups communicate with each other through *intercommunity edges*, a concept called *betweeness of edges*. Once the connection between the individuals in the communities and the interaction between the communities itself is established, then the next step is the placement of the communities in a square location on a grid. This can be represented by S_{pq} (i.e., a square in position of p, q). The next step

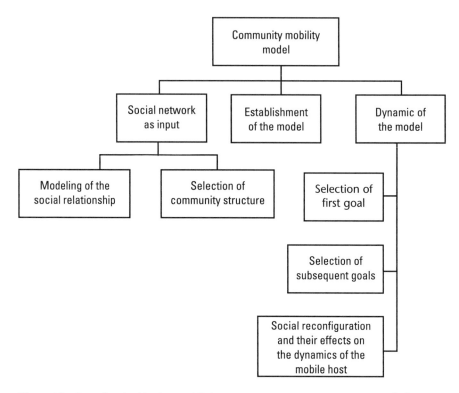

Figure 4.3 Steps involved in the establishment of the community mobility model [25].

is the dynamics of the mobile host. For mobility, a host from each group or community is selected. For each of the hosts the first goal is randomly chosen inside the square S_{pq}. Here, the goal represents the mobility position. The next goal is selected by the *social attractivity*. Each host will have a certain attraction for another host representing another square location. When a host is attracted to another host, then the community moves from the present square location to the square location of another host to which the present host is attracted. Finally, the mobile host needs to be associated with the mobility dynamics.

4.3.11 Semi-Markov Smooth Mobility Model

Semi-Markov smooth (SMS) mobility model obeys the physical law of smooth mobility. It has three phases: (a) speed up phase, (b) middle smooth phase, and (c) slow down phase. In the speed up phase or α phase, the object is accelerated until it reaches an acceptable speed. The speed up phase time interval is given by $[t_0, t_\alpha] = [t_0, t_0 + \alpha \Delta t]$, where t_0 is the initial time and t_α is the alpha phase time. Once the object is moving at an acceptable normal speed, then it has to maintain this speed for some time. The middle smooth phase or β phase is represented by $[t_\alpha, t_\beta] = [t_\alpha, t_\beta + \beta \Delta t]$, where is uniformly distributed over β. As we know, every moving object at some point of time has to slow down and eventually stop. The node reaches γ phase or slow down phase. The node remains in this phase for a random period of time before the speed reaches to $V_\gamma = 0$. Distance between any two points is treated as the Euclidian distance. Trace length is the actual length traveled by a node during one movement without any change in that direction. Distance evolution is the way the traveling of a node is observed. If the speed of a node increases continuously, then the α phase and β phase of the trace length is equivalent to the distance evolution. This results in a smooth trace. When the speed of the node decreases in the γ phase, it results in a sharp curve. SMS mobility model can also be considered a group mobility model. In this scenario a node is selected as a leader. This group leader selects a speed and direction that the other nodes do follow. The velocity of a group member m at its nth step is given by

$$V_n^m = V_n^{Leader} + (1-\rho) \cdot U \cdot \Delta V_{max}$$
$$\phi_n^m = \phi_n^{Leader} + (1 - \rho \cdot U \cdot \Delta \phi_{max})$$

where U is a random variable, ΔV_{max} is the maximum speed, and ϕ_{max} is the difference in the direction of the common node and the leader node in one step [27, 28].

4.4 Simulation Environment

Simulations are performed using information from Section 3.6. The various simulation parameters used are provided in Table 4.1.

Table 4.1
Simulation Parameters

Simulation Time	500s
Simulation area	1000 × 1000m
Number of nodes	30
Transmission range	250m
Mobility model	Community model, GFMM, RPGM, Manhattan model, RWP-SS
Maximum speed	5, 10, 15, 20, 25 (m/s)
Pause time	10 s
CBR sources	15
Data payload	512 Bytes
Traffic rate	5 packets/sec

The performance metrics that are discussed in Section 3.6 including average hop count have been selected for evaluating the mobility models. Average hop count is defined as

$$\frac{\sum CBRnumFwds}{\sum CBRrecv}$$

Each of the group mobility models is mapped against the mobility speed. The mobility speed is varied from 5 m/s to 25 m/s in step 5. All the simulations were run independently and their results were averaged over at five different seeds. AODV routing protocol is used as the underlying routing protocol for simulation. Each of the group mobility models needed different configurations in their file. In community mobility model, as discussed in Section 4.3.10, various communities need to communicate with each other. This is done using a parameter called *rewiring probability*. The value of this parameter was set at 0.1. The number of rows and columns was set at 3. In GFMM, three different force mobility models forming loose or tight groups among the nodes were proposed. For our simulation we have chosen the Fisher-Snedecor or F group force mobility model. For the number of groups' parameter, a value of three was used, so we could have ten nodes in each group. For the RPGM and Manhattan models, the IMPORTANT framework [20] was used to generate the mobility files. For RPGM, the number of groups was specified as three and that amounts to ten nodes in each group. The same maps that were supplied with the source code were used to generate the mobility files. For Manhattan model, the horizontal and vertical parameters were set at 3 and the number of lanes per street was set at 2. RWP-SS did not have much configuration to do as it has fewer parameters.

4.5 Result Analysis

4.5.1 Evaluation of Performance Metrics with Mobility Speed

The packet delivery ratio of community model has the highest packet delivery ratio when compared with other models (Figure 4.4). At 25 m/s the PDR of community model and GFMM are comparable. The delivery ratio of all the mobility models decrease as the mobility speed is increased. The delivery ratio RWP-SS and Manhattan model are comparable. The community model and the GFMM maintain a 90th percentile range even at higher speed, while the other mobility models reduce to 80th percentile range.

The average network delay of RWP-SS is the highest, and it remains at 0.23 no matter at what speed the nodes are moving. As mobility increases, an

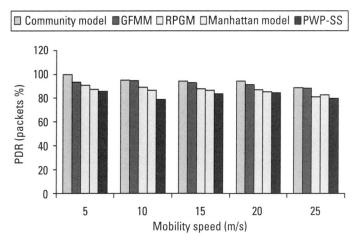

Figure 4.4 PDR vs. mobility speed [29].

extensive link break is observed in the network. A decrease and increase pattern is observed for the community mobility model. There is less delay up to 15 m/s, but as the mobility speed increases then the delay increases, indicating congestion in the network (Figure 4.5).

Results from Figure 4.6 suggest that the throughput performance decreases when the node mobility speed is increased. Community model has the highest throughput of 949 bps when operating at a mobile speed of 5 m/s. The throughput of all the mobility models decrease as the mobility speed is increased. Our results adhere to the analytical results obtained in [1], which states that the "throughput decreases when the transmission range of various mobile nodes cross with each other." In a group mobility model, since the mobile

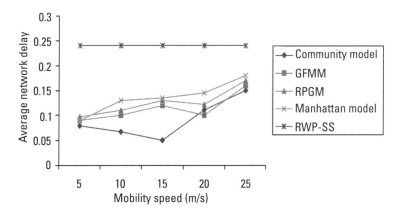

Figure 4.5 Average network delay vs. mobility speed [29].

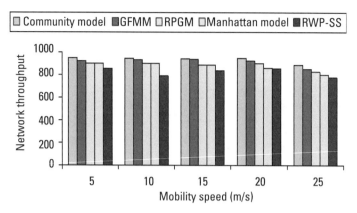

Figure 4.6 Network throughput vs. mobility speed [29].

nodes move strictly in a group, there might be clash of transmission range of various mobile nodes.

From Figure 4.7, the routing overhead of group based mobility models is less than RWP-SS. But, as the mobility speed increases, the overhead of group mobility comes to the vicinity of RWP-SS. In the community model and GFMM, all the nodes are very much packed together in the neighborhood, so there is no need for frequent route discovery with in a group. At high mobility the nodes move frequently, and there is a large chance that many of the nodes get out of the transmission range. This increases the probability of traffic among various intergroup mobile nodes as there is a need to find the nodes for transmission resulting in more overhead.

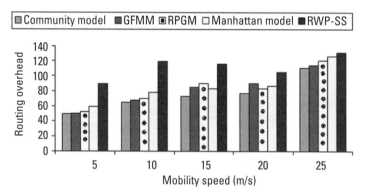

Figure 4.7 Routing overhead vs. mobility speed.

The average hop count (Figure 4.8) of RPGM is less than all the other mobility models. In RPGM, the nodes are grouped together tightly and have less hop count. The number of hops in the community and Manhattan models is more or less equal. This was expected because the community model needs rows and columns parameters to be specified and the Manhattan model has various horizontal and vertical streets for mobility. This explains the same hop count of the community model and Manhattan model.

The energy consumed is the amount of energy consumed by all the nodes at the end of the simulation. The Manhattan model consumes more energy due to less link duration and high spatial dependence of velocity among the nodes. Another generalized reason many be due to high mobility, which results in frequent breakup of routes and changes in topology. Here, it is assumed that all the nodes have same energy level at the beginning of the simulation. We saw a peculiar situation where even though RWP-SS has more routing overhead, the energy consumption is less. This might be due to some nodes that might not move with the same speed as other nodes in RWP-SS, while all the nodes in a group mobility model need to move in the group (Figure 4.9).

4.5.2 Evaluation of Performance Metrics for Different Network Scenarios

A random way point mobility pattern was generated using the scenario generation tool, which supports the format of NS2 distribution. The nodes are configured with a constant pause interval of 100 sec. The speed is a uniform random variable, with a maximum value changing in each simulation run.

Figure 4.10 and Figure 4.11 show the packet delivery ratio of random way point, random walk with reflection, and random walk with wrapping mobility

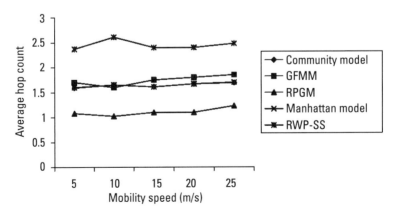

Figure 4.8 Average hopcount vs. mobility speed.

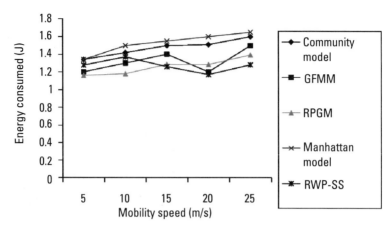

Figure 4.9 Energy consumed vs. mobility speed [29].

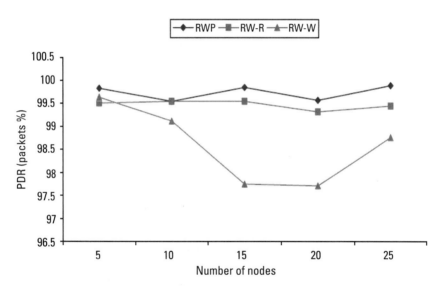

Figure 4.10 PDR vs. number of nodes [30].

models. RWP has the highest packet delivery when compared to RW-R and RW-W with varying mobile nodes. There is a drastic fall in packet delivery in case of RW-W. With varying pause, RW-R scores over RWP. But RW-W packet delivery decreases again.

The delay on the random waypoint is less compared with other two models. We observed that random walk with wrapping has the highest delay with node density as well as with pause time. The relative rankings of protocols may vary with the mobility model used (Figure 4.12 and Figure 4.13).

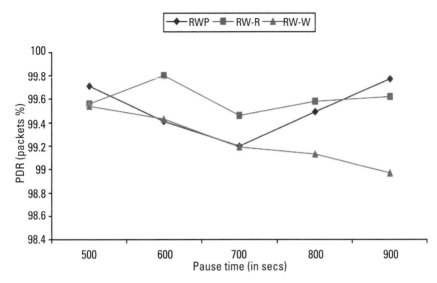

Figure 4.11 PDR vs. pause time [30].

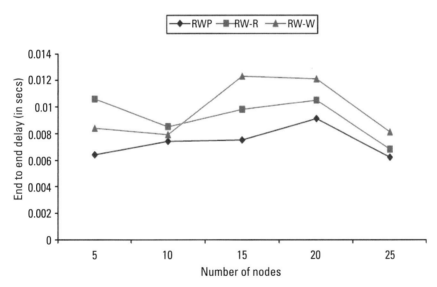

Figure 4.12 End-to-end delay vs. number of nodes [30].

The effect on the routing overhead is much less with the random walk model with wrapping. But the other two models suffer a lot from routing overhead packets. We also observed that random walk with wrapping achieves the lowest routing overhead with node density as well as mobility (Figure 4.14 and Figure 4.15).

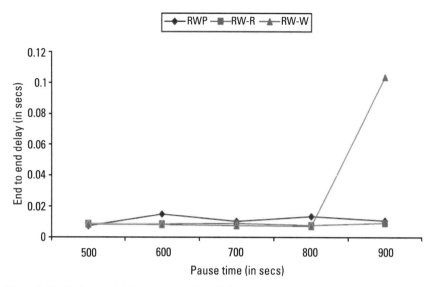

Figure 4.13 End-to-end delay vs. pause time [30].

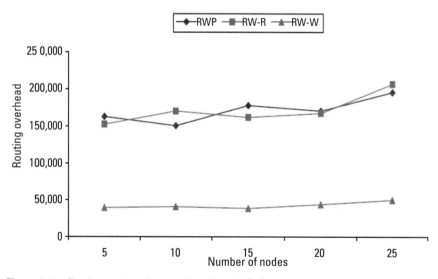

Figure 4.14 Routing overhead vs. number of nodes [30].

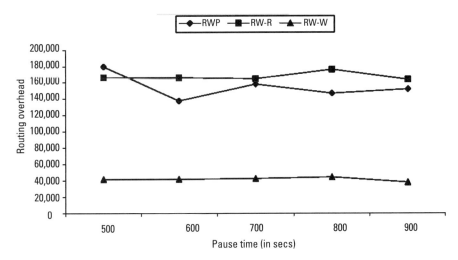

Figure 4.15 Routing overhead vs. pause time [30].

References

[1] Li, X., D. P. Agrawal, and Q.-A. Zeng, "Impact of Mobility on the Performance of Mobile Ad Hoc Network," IEEE International Symposium on Wireless Telecommunications, 2004.

[2] Camp, T., J. Boleng, and V. Davies, "A Survey of Mobility Models for Ad Hoc Network Research," Special Issue on Mobile Ad hoc Networking: Research, Trends and Applications, *Journal of Wireless Communications and Mobile Computing*, 2002, pp. 483–502.

[3] Bai, F., and A. Helmy, "A Survey of Mobility Model," Kluwener Academic Publishers.

[4] Ravikiran, G., and S. Singh, "Influence of Mobility Models on the Performance of Routing Protocols in Ad-Hoc Wireless Networks" IEEE VTC'04 (Spring), Milan, Italy, May 17–19, 2004.

[5] McNeill, K., and Y. Zhao, et al., "Scenario Oriented Ad Hoc Networks Simulation with Mobility Models," MILCOM 2004, IEEE Military Communications Conference, 2004.

[6] Ariyakhajorn, J., et al., "A Comparative Study of Random Waypoint and Gauss Markov Mobility Models in the Performance Evaluation of MANET," International Symposium on Communications and Information Technologies, ISCIT-06, Bangkok, 2006.

[7] Divecha, B., et al, "Analysis of Dynamic Source Routing and Destination Sequenced Distance Vector Protocols for Different Mobility Models." In *Proceedings of the First Asian International Conference on Modelling and Simulation (AMS'07)*, IEEE, 2007.

[8] Djenouri, D., E. Nekka, and W. Soualhi, "Simulation of Mobility Models in Vehicular Ad Hoc Networks." In *Proceedings of Ambi-Sys Workshop on Software Organization and Monitoring of Ambient Systems*, Quebec City, Canada, February 2008.

[9] Harri, J., F. Filali, and C. Bonnet, "Mobility Models for Vehicular Ad Hoc Networks: A Survey and Taxonomy," *Research Report RR-06-168*, Institute Eurecom, Department of Mobile Communications, Sophia-Antipolis, France.

[10] Greede, A., S. M. Allen, and R. M.Whitaker, "A Simple Human Mobility Model for Opportunistic Networks," The 9th Annual Postgraduate Symposium, The Convergence of Telecommunications, Networking, and Broadcasting, John Moores University, Liverpool, June 23–24, 2008.

[11] Jardosh, A., E. Royer, et al., "Towards Realistic Mobility Models for Mobile Ad hoc Networks." In *Proceedings of IEEE MobiCom*, San Diego, CA, September 2003.

[12] Lassila, P., E. Hyytia, and H. Koskinen, "Connectivity Properties of Random Waypoint Mobility Model for Ad Hoc Networks," Springer Science+Business Media. Reprinted with permission from *Proceedings of the Fourth Annual Mediterranean Workshop on Ad Hoc Networks (Med-Hoc-Net)*, June 2005.

[13] Johnson, D. B., and D. A. Maltz, "Dynamic Source Routing in Ad Hoc Wireless Networks." In *Ad Hoc Networking*, ed. Charles E. Perkins, Reading, MA: Addison Wesley, Chapter 5, 2001, pp. 139–172.

[14] Bettstetter, C., and C. Wagner, "The Spatial Node Distribution of the Random Waypoint Mobility Model," Mobile Ad Hoc Netzwerke, 1. Deutcher Workshop uber Mobile Ad-hoc Netzwerke WMAN 2002.

[15] Navidi, W., T. Camp, and N. Bauer, "Improving the Accuracy of Random Waypoint Simulations Through Steady-State Initialization." In *Proceedings of the 15th International Conference on Modeling and Simulation (MS'04)*, March 2004, pp. 319–326.

[16] Le Boudec, J.-Y., and M. Vojnovic, "Perfect Simulation and Stationary of a Class of Mobility Models." In *Proceedings of IEEE INFOCOM 2005*, Miami, FL, March 13-17, 2005.

[17] Liang, B., and J. H. Zygmunt, "Predictive Distance Based Mobility Management for PCS Networks." In *Proceedings of IEEE INFOCOM-99*, New York, March 21–25, 1999.

[18] BonnMotion Software, http://net.cs.uni-bonn.de/wg/cs/applications/bonnmotion, last accessed: 2009–2010.

[19] Hong, X., et al, "A Group Mobility Model for Ad Hoc Wireless Networks." In *Proceedings of the ACM/IEEE MSWIM'99*, Seattle, WA, August 1999, pp. 53–60.

[20] Bai, F., N. Sadagopan, and A. Helmy, "The IMPORTANT Framework for Analyzing the Impact of Mobility on Performance of Routing for Ad Hoc Networks," *Ad Hoc Networks Journal*, Vol. 1, No. 4, November 2003, pp. 383–403.

[21] Williams, S. A., and H. Dijiang, "A Group Force Mobility Model," 9th Communications and Networking Simulation Symposium, April 2–6, 2006.

[22] Meghanathan, N., "Review of Mobility Models," Department of Computer Science, Jackson State University, Jackson, MS.

[23] Rhee, I., M. Shin, et al., "On the Levy-Walk Nature of Human Mobility," INFOCOM, Arizona, 2008.

[24] Hong, S., and I. Rhee, et al., "Routing Performance Analysis of Human Driven Delay Tolerant Networks Using the Truncated Levy Walk Model," ACM SIGMOBILE

International Workshop on Mobility Models for Networking Research (Colocated with MobiHoc 08), Hong Kong, 2008.

[25] Musolesi, M., and C. Mascolo, "A Community Based Mobility Model for Ad Hoc Network Research," International Symposium on Mobile Ad Hoc Networking and Computing, *Proceedings of the 2nd International Workshop on Multi Hop Ad Hoc Networks*, 2006.

[26] Newman M. E. J., and M. Girvan, "Finding and Evaluating Community Structure in Networks," *Physical Review*, Februrary 2004, Vol. 69.

[27] Zhao, M., and W. Wang, "A Novel Semi Markov Smooth Mobility Model for Mobile Ad Hoc Networks." In *Proceedings of the IEEE GLOBECOM'06*, San Francisco, CA, November 2006.

[28] Zhao, M., and W. Wang, "Design and Applications of a Smooth Mobility Model for Mobile Ad Hoc Networks." In *Proceedings of the IEEE Milcom'06*, Washington DC, October 2006.

[29] Proceedings of the First IEEE International Conferenece on Networks and Communications, December 27–29, 2009, Chennai, India.

[30] *International Journal of Computer Science and Network Security*, Vol. 7, No. 6, June 2007.

5

Overhead Control Mechanism and Analysis in MANET

5.1 Introduction

Wireless networks provide users with seamless access to information like checking email, browsing the Internet, and accessing corporate data without getting disconnected from the network even when the user is in motion. Wireless technologies have enabled us to lead a convenient way of life by becoming part and parcel of our daily routine. Mobile ad hoc networks are autonomous systems formed by mobile nodes without any infrastructure support. Routing in MANET is challenging because of the dynamic nature of the network topology. For fixed network routing protocols it can be assumed that all the routers have a sufficient description of the underlying network, either through global or decentralized routing tables. But dynamic wireless networks do not easily admit such topology state knowledge. A minimum overhead is associated with the routing in MANET and is greater than that of fixed networks because of the inherent randomness of the topology. Therefore, it is of interest to know how small the routing overhead can be made for a given routing strategy and random topology [1].

  We have made a substantial effort to study the performance of various routing protocols, like AOMDV, TORA, and OLSR, under three different modles: the Bansal energy model, the Vaddina energy model, and the Chandrakasan energy model. We have also made an unassailable effort to study the energy consumption pattern of five mobility models under five different routing protocols.

Several performance metrics such as packet delivery ratio, average end-to-end delay, and routing overhead are commonly used to evaluate the performance of routing protocols in MANET. Metrics routing overhead is the important one among these, as it determines the scalability of a routing protocol. Routing overhead dictates how many extra messages were used to achieve the acceptance rate of improvement.

In mobile ad hoc routing protocols, different methods are followed to evaluate the routing overhead. Some of them are physical experiments, simulations, and theoretical analysis [2].

In simulations a controlled environment is provided to test and debug many of the routing protocols. Therefore, most of the literature [3–5] evaluates the control overhead in routing protocols using software simulators like NS2, Glomosim, Qualnet, and OPNET. However, simulations are not foolproof and may fail to accurately reveal some critical behaviors of routing protocols, as most of the simulation experiments are based on simplified assumptions.

Physical experiments determine the performance of routing protocols by implementing them in a real environment. Some of the investigations in the literature evaluate routing overhead in a real physical environment [6, 7]. But physical experiments are much more difficult and time consuming to be carried out. Analysis of routing overhead from a theoretical point of view provides a deeper understanding of advantages, limitations, and tradeoffs found in the routing protocols in MANET.

In this chapter, we study the energy overhead performance of three different routing protocols under three different energy models. The three different energy models considered are (a) Bansal energy model, (b) Vaddina energy model, and (c) Chandrakasan energy model. We employ these energy models to the AOMDV, TORA, and OLSR routing protocols to determine the energy overhead among these three routing protocols by varying the transmission range, and the prime aim is not to determine which energy model has less overhead, but to analyze how these routing protocols behave under different energy model.

We have analyzed the energy consumption of five mobility models under five different routing protocols. The mobility models considered are random walk with wrapping (RW-W), random waypoint with steady state (RW-PSS), Gauss-Markov mobility model (GM), community based mobility model (CM), and semi-Markov smooth mobility model (SMS). The selected routing protocols are AOMDV, DYMO, FSR, OLSR, and TORA. In the literature many have concentrated on the amount of energy consumed by the routing protocols. But there is no detailed analysis of the mobility model's role on routing protocols in energy consumption. We also found contradicting results for the TORA routing protocol. The energy consumed by each of these mobility models are mapped by varying the mobility speed. From our simulation we deduce that with increased complexity of the mobility models, the amount of energy consumed also increases among the routing protocols.

To the best of our knowledge, no work has been reported that compares and studies the energy overhead of these three routing protocols under three different energy models. Also no work has been reported that compares and studies the performance of all these five mobility models under five different routing protocols. The results obtained for our work are in complete conjugation with the results obtained in [8].

5.2 Literature Background

Early work on energy consumption in ad hoc networks was done by Feeney et al. [9]. The authors conduct various experiments to determine the energy consumption of a Lucent Wireless WaveLan IEEE 802.11 network card. The authors also formulate a linear equation to quantify the *per packet energy consumption*.

Preetha Prabhakaran et al. [10] considered various mobility models like random waypoint (RWP), Manhattan grid model (MG), Gauss-Markov model (GM), community mobility model (CM), and RPGM and analyzed the energy consumption to transmit, receive, and drop the control packets. They have computed the energy consumption by mapping it against the mobility speed. They have shown that as the mobility speed is increased the energy goodput also decreases. Among these three mobility models, RWP has the highest energy consumption. For CM and RPGM mobility models, it is shown that as the number of groups increases then the energy consumption also decreases.

In Juan-Carlos Cano et al. [8], the authors have mapped the energy consumption of AODV, DSDV, DSR, and TORA routing protocols. The authors have calculated the energy consumed by these four routing protocols by mapping them against varying mobility speed, traffic patterns, node numbers, and area. The authors conclude that the TORA routing protocol had the worst performance in all the scenarios.

In Bor-rong Chen et al. [11] as in [8], three different routing protocols (AODV, DSR, and DSDV) are considered. They are compared against random waypoint, RPGM, and Manhattan grid models. They use the same energy model specified by Feeney et al. [9]. Through simulation the authors show that AODV has more energy consumption under RWP and RPGM, while DSR consumes more energy under the Manhattan grid model.

In [8] the energy analysis of four routing protocols (AODV, DSR, DSDV, and TORA) is carried out. The energy consumption of these routing protocols is studied with varying mobility, different dimensions of area, node density, number of sources, and the traffic pattern. At different mobile speeds and traffic loads, the DSDV routing protocol has less energy consumption. On varying the node density, AODV and DSR have less energy consumption. When different dimensions are considered, TORA has the worst performance while the remaining protocols more or less have the same performance.

The effect of routing protocols on mobility models and the effect of mobility models on routing protocols is carried out in [11]. The routing protocols considered are AODV, DSDV, and DSR. The underlying mobility models are random waypoint, Manhattan grid, and RPGM. Under random waypoint DSR has the least energy consumption. For the Manhattan grid mobility model at lower speed DSR consumes less energy but as the mobility speed is increased then DSDV shows better performance. AODV and DSR routing protocols score over DSDV by consuming less energy for the RPGM mobility model.

The energy consumption of the TORA routing protocol is studied under random waypoint and reference point group mobility by varying the mobility speed [12]. Through simulation it is shown that the UDP traffic consumes less energy than that of TCP traffic for RPGM mobility model.

5.3 Overhead Analysis in Hierarchical Routing Scheme

The routing schemes for ad hoc networks are classified into proactive and reactive routing protocols. Proactive protocols like OLSR and DSDV maintain routing information about the available paths in the network even if these paths are not currently used. Such paths have the drawback that it may occupy a significant part of the available bandwidth. Reactive routing protocols like AODV, TORA, and DSR maintain only the routes that are currently available. However, they still generate large amount of control traffic when the topology of network changes frequently.

Therefore, many of the routing protocols become inherently not scalable with respect to the number of nodes and the control overhead because of the properties of frequent route breakage and unpredictable topology changes in MANET. A hierarchy of layers is usually imposed on the network to provide routing scalability. In ad hoc networks scalability issues are handled hierarchically. Many hierarchical routing algorithms like cluster based routing and the dominating set based routing are adopted for routing in ad hoc wireless networks.

Sucec and Marsic suggest a formal analysis of the routing overhead by providing a theoretical upper bound on the communication overhead incurred by the clustering algorithms that adopt the hierarchical routing schemes. There are many scalability performance metrics like routing table storage overhead, least hop path length, and hierarchical path length. Out of these metrics, control overhead per node (Ψ) is the most important one because of scarce wireless link capacity, which has severe performance limitation. The control overhead Ψ is expressed as a function of V, the set of network nodes. With reasonable assumptions it can be established that the average overhead generated per node per second is only polylogarithmic in the node count (i.e., $\Psi = O(\log^2|V|)$ bits per second per node [13]).

In hierarchically organized networks, communication overhead may result from the following phenomenon: (a) hello protocols, (b) level-k cluster formation and cluster maintenance messaging, $k \in \{1,2...L\}$, where L is the number of levels in the clustered hierarchy, (c) flooding of the cluster topology updates to cluster members, (d) addressing information required in datagram headers, (e) location management events due to changes in the clustered hierarchy and due to node mobility between clusters, (f) hand off or transfer of location management data, and (g) location query events. In hierarchically organized networks, total communication overhead per node ψ is the sum of these contributing elements.

The control overhead and network throughput under a cluster based hierarchical routing scheme are discussed [14]. In a hierarchical design, it is claimed that when the routing overhead is minimized, then there is a loss in the throughput from the same hierarchical design. A strict hierarchical routing that is not based on any specific routing protocol is considered,. Hierarchical routing protocols in MANET do not require every node to know the entire topology information. Only a few nodes, called the *cluster head nodes*, can simply send their packets to these cluster heads. The cluster head nodes need to know about the entire topology information and all other nodes can simply send their packets to these cluster heads.

Hierarchical routing protocols reduce the routing overhead, as lesser nodes need to know the topology information of an ad hoc network. With hierarchical routing scheme the throughput of ad hoc network is smaller by a factor of $O\left(\dfrac{N2}{N1}\right)$, where $N2$ is the number of cluster head nodes and $N1$ is the number of all the nodes in the network. Therefore, it is said that there is a tradeoff between the gain from the routing overhead and the loss in the throughput from the hierarchical design of the ad hoc routing protocols.

The control overhead in a hierarchical routing scheme can be due to the address management or location management, due to the maintenance of routing tables as well as due to packet transmissions per node per second (ϕ). Hence the overhead ϕ required by hierarchical routing is a polylogarithmic function of the network node count (N) (i.e., $\Phi = \Theta(\log^2 |N|)$ packet transmissions per node per second). In this equation, overhead due to hierarchical cluster formation and location management are identified [15].

5.4 Overhead Minimizing Techniques and Analysis Using Clustering Mechanisms

Different clustering schemes may use different clustering algorithms. However, generally three different types of control messages are needed: (a) beacon messages, also known as hello messages, are used to learn about the identities of the neighboring nodes, (b) cluster messages are used to adapt to cluster changes and to update the role of a node, and (c) route messages are used to learn about the possible route changes in the network [16].

Different types of control messages overhead are as follows:

1. *Hello overhead:* To reduce the hello overhead messages, the frequency of hello messages generated by a node to learn about its neighboring node when a new link is formed should be at least equal to the link generation rate. The link generation between any two nodes can be notified by sending the hello messages, and each of the nodes can hear the hello message sent by the other node.

2. *Cluster message overhead due to link break between cluster members and their cluster heads:* This event causes the node to change its cluster or become a cluster head when it has no neighboring clustering heads. The cluster members send the cluster messages due to this type of link change. To minimize the control message overhead, the ratio of such link breaks to total link breaks should be equal to the ratio of links between the cluster members and cluster heads divided by the total number of links in the entire network.

3. *Cluster message overhead due to link generation between two cluster heads:* When a link is generated between two cluster heads, one of the cluster heads needs to give up its cluster head role, which is decided by the clustering algorithm. Every time a link between two cluster heads appears, the number of cluster messages generated is same as the number of nodes in the cluster that needs to undergo reclustering.

4. *Routing overhead:* When a particular node in the cluster should be updated with the route to other nodes in the cluster, the routing storage overhead is proportional to the size of the cluster.

For mobile ad hoc networks, an analytical study on routing overhead of two level cluster based routing protocols is performed. Routing protocols are summarized into generic proactive routing protocol and a generic reactive routing protocol. It's generic because there may be some different strategy employed for each of the groups, but the underlying nature of the routing is similar.

For two-level cluster based routing scheme, the routing is divided into two separate parts—routing among different clusters (i.e., intercluster routing) and routing within a cluster (i.e., intracluster routing). Since there are two types of routing schemes—proactive and reactive—that can be employed in intercluster routing and intracluster routing, there are four types of two-level hierarchical routing schemes with different combinations of them. So we have proactive to proactive, reactive to reactive, proactive to reactive, and reactive to proactive routing schemes.

For intercluster routing, when a proactive scheme is adapted each cluster head periodically collects its local cluster topology and then broadcasts it to its direct neighbor cluster head via gateways. When a reactive routing scheme is used for intercluster routing, a route request for a route to the destination node that is in another cluster will be broadcasted among cluster heads. When a proactive routing scheme is utilized for intracluster routing, each node broadcasts its local node topology information, so the route to the destination within the same cluster will be available when needed. When a reactive routing scheme is employed for intracluster routing, a route request to the destination will be flooded within the same cluster.

A proactive to proactive routing scheme thus will work as a proactive routing protocol with a hierarchical structure. The proactive-to-proactive routing scheme produces overhead due to periodical topology maintenance of

$$O\left(\frac{1}{n}N^2 + \frac{1}{kN_c}N^2\right)$$

where n is the total number of clusters in the network, N is the network size, k is the cluster radius, and N_c is the cluster size.

Further, a reactive-to-reactive routing protocol operates as a purely reactive routing protocol with a hierarchical structure. A reactive-to-reactive routing protocol yields a routing overhead due to route discovery of

$$O\left(\frac{1}{k}N^2\right)$$

The cluster heads periodically exchange topological information, and a node always sends a route request to its cluster head where there is no available route to an expected destination in a proactive-to-reactive routing scheme. Then the cluster head will send a route reply packet to the node, which indicates that the destination is within the local cluster or contains a route to the destination node, which is in another cluster. Proactive-to-reactive routing protocols have a basic routing overhead due to topology maintenance, cluster maintenance, and route discovery, which is found to be

$$O\left(\frac{1}{n}N^2 + \frac{r}{nk}N^2\right)$$

where r is average route lifetime.

In a reactive-to-proactive routing scheme each node in the cluster will periodically broadcast local node topology information within the cluster. Thus, when the destination is within the cluster, the route is immediately available. Otherwise, the node will send a route request packet to its cluster head, which will be broadcasted among the cluster heads. Reactive to proactive routing protocol have a basic routing overhead due to cluster maintenance and route discovery, which is approximately equal to

$$O\left(\frac{1}{n}N^2 + \frac{1}{k}N^2\right)$$

A mathematical framework for quantifying the overhead of a cluster based routing protocol (D-hop max min) is investigated by Wu and Abouzeid [17]. The authors provide a relationship between routing overhead and route request traffic pattern. From a routing protocol perspective, *traffic* could be defined as the pattern by which the source-destination pairs are chosen. The choice of a source-destination pair depends on the number of hops along the shortest path between them. Also the network topology is modeled as a regular two-dimensional grid of unreliable nodes. It is assumed that an infinite number of nodes are located at the intersections of a regular grid. The transmission range of each node is limited such that a node can directly communicate only with its four immediate neighbors. It is reported that the clustering does not change the traffic requirement for infinite scalability compared to flat protocols, but reduces the overhead by a factor of

$$O\left(\frac{1}{M}\right)$$

where M is the cluster size.

Wu and Abouzeid show that the routing overhead can be attributed to events like (a) route discovery, (b) route maintenance, and (c) cluster maintenance.

Route discovery is the mechanism whereby a node i wishing to send a packet to the destination j obtains a route to j. When a source node i wants to send a packet to the destination node j, it first sends a route request packet to its cluster head along the shortest path. The route reply packet travels across the shortest path back to the cluster head that initiated the RREQ packet. So the route discovery event involves an RREP and RREQ processes. The overhead for RREQ is generally higher than the RREP since it may involve flooding at the cluster head level. Therefore, the minimum average overhead of finding a new route is

$$\frac{f(k-3)\left(6M+6+\dfrac{3}{M}\right)+f(k-2)(4M^2+2)-f(k-1)M(M^2-1)}{3(M^2+2M+1)f(k-1)}$$

where M is the radius of the cluster and changing the value of k controls the traffic pattern.

Route maintenance is the mechanism by which a node i is notified that a link along an active path has broken such that it can no longer reach the destination node j through that route. When route maintenance indicates a link is broken, i may invoke route discovery again to find a new route for subsequent packets to j. In cluster based routing, the neighboring node sends a RERR packet to its cluster head rather than the source node itself. The cluster head could patch a path locally without informing the source node, if the failed node is not the destination node. This is called *local repair*. In this case, the path is locally fixed. Also, the RERR packet sent from a neighboring node to the cluster head is considered as the cluster maintenance overhead. Therefore the minimum overhead required for route maintenance is $4C(f(k-2) - f(k-1))$ where $k > 3$ and C is a constant.

Clustering incurs cluster maintenance overhead, which is the amount of control packets needed to maintain the cluster membership information. The membership in each cluster changes over time in response to node mobility, node failure, or new node arrival. The average cluster maintenance overhead is

$$\frac{4(M-1)M(M+1)}{3(2M^2+2M+1)}+2M$$

where M is the radius of the cluster.

Zhou [18] provides a scalability analysis of the routing overheads with regard to the number of nodes and the cluster size. Both the interior routing overhead within the cluster and the exterior routing overhead across the clusters are considered. The routing protocol includes a mechanism for detecting, collecting, and distributing the network topology changes. The process of detecting, collecting, and distributing the network topology changes contributes to a total routing overhead R_t. The routing overhead can be separated into interior routing overhead (R_i) and the exterior routing overhead (R_e). The interior routing overhead R_i is the bit rate needed to maintain the local detailed topology. This includes the overhead of detecting the link status changes by sending hello messages, updating the cluster head about the changes in link status, and maintaining the shortest path between the regular nodes to their cluster head.

In order to maintain the global ownership topology, exterior routing overhead (R_e) is the bit rate needed, which includes the overheads of the distributing the local ownership topologies among the cluster heads. Hence $R_t = R_i + R_e$.

5.5 Overhead Minimization by Header Compression

In the literature [19], we can find in most of the studies approximately half of the packets sent across the Internet are 80 bytes long or less. This percentage has increased over the last few years in part due to widespread use of real-time multimedia applications. The multimedia application's packet is usually smaller in size, and these small packets must be added with many protocol headers while traveling through the networks. In IPv4 networks there can be at least 28 bytes (UDP) or 40 bytes (TCP) overhead per packet. This overhead consumes much of the bandwidth, which is very limited in wireless links. Small packets and relatively larger header size translates into poor line efficiency. Line efficiency increases with the increase in the amount of actual data transmitted. Ad hoc networks create additional challenges such as context initialization overhead and packet reordering issues associated with node mobility. The dynamic nature of ad hoc networks has a negative impact on header compression efficiency. Because of the dynamic topology of MANET, routes get broken and frequently we are to reestablish new routes. Then the node has to examine the header of the packet. If the header is compressed, it must be decompressed to examine the content of the header. This decompression process will have impact on computation, which degrades the performance of the MANET.

A context is established by first sending a packet with a full uncompressed header that provides a common knowledge between the sender and receiver about the static field values as well as the initial values for dynamic fields. This stage is known as context initialization. Then the subsequent compressed headers are interpreted and decompressed according to a previously established context. Every packet contains a context label. Here, the context label indicates the context in which the headers are compressed or decompressed.

A novel hop-by-hop context initialization algorithm proposed by Jesus et al. [20] depends on the routing information to reduce the overhead associated with the context initialization of IP headers and uses a stateless compression method to reduce the overhead associated with the control messages. Context initialization of IP headers is done on a hop-by-hop basis because the headers need to be examined in an uncompressed state at each of the intermediate nodes. The context initialization overhead is reduced by caching the address information that is transmitted in the routing messages in order to reduce the size of the context initialization headers.

Also, a stateless header compression is proposed. It is stateless because the state of the context is fixed and does not change with time. Header compression improves the line efficiency by exploiting the redundancies between the header fields within the same packet or consecutive packets belonging to the same stream.

The overall result is the reduced overhead, increased network capacity, and line efficiency, even in the presence of rapid path fluctuations.

An ad-hoc robust header compression (ARHC) [21] protocol can be used to compress UDP, TCP, and raw IP headers in ad hoc network. The mechanism of ARHC is that when the first packet of a session arrives, the compressor generates a unique number called context ID, which indexes the quintuplet (source address, destination address, source port, destination port, protocol) and all the constant fields. The compressor then records the context id, quintuplet, and all the constant fields. Then the compressor will send the full packet header, along with the context ID. Upon receiving the very first packet, the decompressor records this information. When the subsequent packets arrive later, the compressor and decompressor act as follows: The compressor will remove the constant fields and the quintuplet from the header and transmits only the context ID. The decompressor then retrieves the quintuplets and the constant fields from the context tables indexed by the context ID, thereby restoring the original header.

5.6 Overhead Minimization for Ad Hoc Networks Connected to the Internet

The Internet has become the backbone of the wired and wireless communication. Also, mobile computing is gaining popularity. In order to meet the rapid growing demand of mobile computing, many researchers are interested in the integration of MANET with the Internet.

When a mobile node in MANET wants to exchange packets with the Internet, first the node must be assigned a global IP address and then the available

Internet gateways have to be discovered to connect to the Internet as shown in Figure 5.1. But, this is achieved at the cost of higher control overhead.

For gateway discovery, a node depends on periodic gateway advertisement. To make efficient use of this periodic advertisement, it is necessary to limit the advertisement flooding area. In a complete adaptive scheme to discover Internet gateway (IG) in an efficient manner for AODV, both the periodic advertisement and adaptive advertisement schemes are used [22]. At a relatively long interval, each gateway sends the periodic advertisement messages. Periodic advertisements performed at a widely spaced interval do not generate a great deal of overhead but still provides the mobile nodes with a good chance of finding the shortest path to a previously used gateway. The TTL of the periodic gateway message is used as a parameter to adjust the network conditions. A heuristic algorithm called *minimal benefit average* [23] decides the next TTL to be used for the periodic gateway advertisement messages.

The goal of the adaptive advertisement scheme is to send advertisement packets only when the gateway detects the movement of nodes, which would result in the paths used by the source mobile nodes communicating with the gateway to be changed. Adaptive advertisement is performed when needed, regardless of the time interval used for periodic advertisement. In this approach there is reduction in overhead messages since the periodic advertisements are sent at a long time interval and perform adaptive advertisement only if there is mobility in the network.

The various parameters that affect the control overhead created by interoperating the ad hoc routing protocols and IP based mobility management protocols is already addressed in some recent works [24]. Mobile IP is used as the baseline mobility management protocol, and AODV is chosen as the ad hoc routing protocol. IP tunneling is used to separate the ad hoc network from the

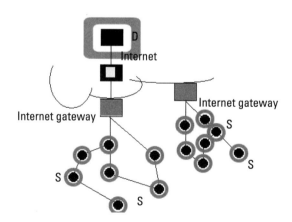

Figure 5.1 MANET connected to the Internet scenario [21].

fixed network. In mobile IP, a mobile node can tell which router is available by listening to router advertisements, which are periodically broadcasted by the routers.

A fixed access router is assigned the role of mobility agent and has a connection to at least one of the MANET nodes. Such a router is referred to as an *ad hoc Internet access router* (AIAR), and it maintains a list called an *ad hoc list*, which keeps a list of IP address of the mobile nodes that wish to have Internet connectivity. In an integrated network, the control overhead comprises of AIAR registration packets, routing protocol control packets, mobile IP registration packets, and mobile IP router advertisement.

In mobile IP, a majority of the overhead is due to the route advertisement packets that are being repeatedly and periodically forwarded among the mobile nodes. Also, the router advertisement used by the mobility management protocol to carry network information is the major source of unnecessary control overhead within MANET. Varying the TTL value is an effective mechanism to control the amount of advertisement packets.

A multihop router is developed [25] for nonuniform route connectivity with low routing overhead. To achieve efficient route discovery and route maintenance, a new routing scheme called the hop-count based routing (HBR) protocol is developed. HBR is an on-demand routing protocol. When a source node needs to discover a route to a node on the wired network, it initiates a route discovery to the nearest access point by broadcasting a route request packet. By utilizing the hop counts referring to access points, the route discovery region is localized to a small area. On receiving the route request packet, the access point responds by sending a route reply packet to the source node. Once a route is found, the source node begins to forward data to the access point. After the access point receives these data packets from the source node, it then forwards these packets through the wired line to the destination node on the wired network using the routing protocol in the wired network. By using the hop count information, an attempt is made to reduce the number of nodes to whom the route request is propagated. Thus, in the HBR routing protocol, constructing routes from mobile nodes to access points hop count information is utilized to localize the route discovery within a limited area in order to reduce the routing overhead.

5.7 Energy Models

Transmission and receiving of energy can be modeled as $E_{(ptx/rcv)} = i * v * t_p$ Joules, where i is the current value, v is the voltage, and t_p is the time taken to transmit or receive the packet [9, 10]. Table 5.1 gives the various values considered for Bansal energy model (BEM) [26], Vaddina energy model (VEM) [27],

Table 5.1
Various Values of Different Energy Models

Energy Model	Transmission Power	Receiving Power	Idle Power
Bansal model	0.0271	0.0135	0.00000739
Vaddina model	0.0744	0.0648	0.00000552
Chandrakasan model	0.175	0.175	0.00000175

and Chandrakasan energy model (CEM) [28]. These energy values are used to obtain the results of Figures 5.2, 5.3, and 5.4.

The amount of energy spent in transmitting and receiving the packets is calculated by using the following equations:

$$Energy_{tx} = (330 * PacketSize)/2 * 10^6$$
$$Energy_{rx} = (230 * PacketSize)/2 * 10^6$$

where packet size is specified in bits.

In the previous energy equations, where 330 is energy consumption 330mA in transmitting mode using 5V energy supply and 230 is energy consumption 230mA in receiving mode using 5V energy supply and is divided by channel bandwidth of 2 Mbps. The average energy consumed equations are applied to analyze the energy consumption of various mobility models under different routing protocols.

5.8 Simulation

The simulation is performed as per Section 3.6. The results from Figures 5.2–5.4 are obtained by using simulation parameters mentioned in Table 5.2, while the results from Figures 5.5–5.9 are through the Table 5.3 simulation parameters. Each of the energy overhead models is mapped against the different transmission ranges varied from 250m to 550m in steps of 50. Here, the idea is not find an optimized transmission or the amount of energy consumed by each protocol. The intention is to see the behavior of these protocols under various energy models. Besides running independently, all the simulations are averaged for five different seeds. The mobility speed is varied from 5 m/s to 25 m/s in steps of five for different mobility models.

Table 5.2
Simulation Parameters for Results from Figure 5.2–Figure 5.4

Simulator	NS2
Routing protocols	AOMDV, TORA, OLSR
Simulation time	500s
Simulation area	1000 x 1000m
Number of nodes	50
Transmission range (m)	250, 300, 350, 400, 450, 500, 550
Mobility model	RWP-SS
Maximum speed	10 (m/s)
Pause time	10 sec
CBR sources	25
Data payload	512 bytes
Traffic rate	5 packets/sec

Table 5.3
Simulation Parameters for Results of Figure 5.5–Figure 5.9

Simulator	NS2
Routing protocols	AOMDV, DYMO, FSR, OLSR, and TORA
Mobility model	Random way point with wrapping, random way point with steady state, Gauss-Markov mobility model, community mobility model, and SMS mobility model
Simulation time (sec)	900
Pause time (sec)	10
Simulation area (m)	1000 x 1000
Number of nodes	50
Transmission range	250m
Maximum speed (m/s)	5, 10, 15, 20, 25
Traffic rate (pkts/sec)	5
Data payload (bytes)	512

5.9 Result Analysis

5.9.1 Performance Evaluation of Different Energy Models

RWP-SS mobility model enables us to compute the results right from starting the simulation time. Conserving energy leads to extended battery life in ad hoc networks. Nodes in the ad hoc networks are battery powered. A number of factors shape the extendibility of battery life in ad hoc networks, like the speed at which the nodes are moving, the transmission range, the amount of packets sent and received, and the amount of information that needs to be processed. It is desirable to find an optimum transmission range to conserve energy without compromising on the amount of data delivered.

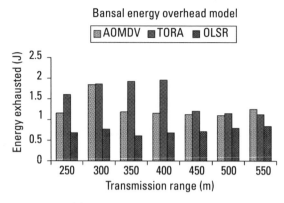

Figure 5.2 Energy exhausted (J) vs. transmission range for BEM [30].

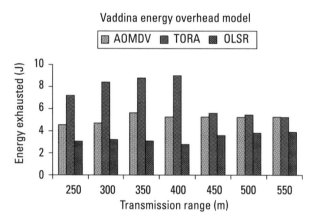

Figure 5.3 Energy exhausted (J) vs. transmission range for VEM [30].

Transmission range plays a very important role in deciding the amount of energy overhead needed for establishing connectivity among various nodes in the network. Increasing the transmission range leads to less hop count, and there will be fewer breaks in the connectivity of the mobile nodes.

The TORA routing protocol has maximum energy consumption at 400m across all the energy models. AOMDV has highest energy consumption at 300m, 350m, and 400m for BEM, VEM, and CEM models. OLSR has peak energy consumption at 550m.

AOMDV maintains high connectivity even at high mobility due to multiple paths resulting in less energy overhead for maintaining the network. Even though AOMDV by virtue of its multiple route maintenance has less energy overhead, the maintenance of these multiple paths itself may lead to energy overhead. The amount of energy spent in signaling tends to decrease with an increase in transmission range across all the routing protocols. Energy

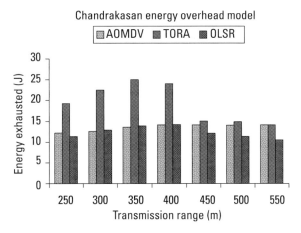

Figure 5.4 Energy exhausted (J) vs. transmission range for CEM [30].

consumption is less in OLSR when compared to TORA and AOMDV. OLSR has less energy overhead, as only MPR selector nodes will broadcast its status across the network.

In TORA the number of nodes that can be accessed increases with the increase in transmission range. This increases the amount of interference and collisions, resulting in retransmission of packets. Energy spent in signaling is maximum for TORA protocol at up to 400m. After this point, the energy consumption is comparable to OLSR and AOMDV routing protocols. This can be summed to the amount of packets used in maintaining links among various nodes in the network. Establishment of connectivity across the nodes is stabilized with the increase in transmission range. This reduces the energy consumed in the network.

In literatures, it has been observed that the packet delivery increases with an increase in transmission range but with higher energy consumption for every transmission [29]. But energy consumption increases up to a transmission range. After that, energy consumption decreases and it remains the same across all the transmission range. This tendency can be seen across all the routing protocols.

When the transmission power and receiving power is less, then there is huge difference in the amount of energy spent in signaling between the AOMDV and OLSR routing protocols, as can be seen from Figures 5.2, 5.3, and 5.4. But the difference starts to decrease with the increase in transmission power and receiving power. In the CEM model, the energy spent in signaling is almost same for both the AOMDV and OLSR routing protocols. After a transmission range of 400m, the amount of energy spent in signaling between the AOMDV and TORA routing protocols remains the same across all the energy models.

5.9.2 Performance Evaluation of Energy Consumed with Mobility Speed

The AOMDV routing protocol consumes more energy under community based mobility mode (Figure 5.5). The energy consumption is steady for AOMDV protocol at 166 to 168 joules when the community model is used. In our simulation we have three groups. Network connectivity has to be maintained for these groups. Since AOMDV is a multipath algorithm, we have many paths running between the nodes of different groups, resulting in higher energy consumption. The Gauss-Markov mobility model comes in at a near second to community model. SMS mobility model has the highest impact on the OLSR routing protocol, followed by the community based mobility model (Figure 5.6). At 25 m/s all the mobility models except the community based mobility

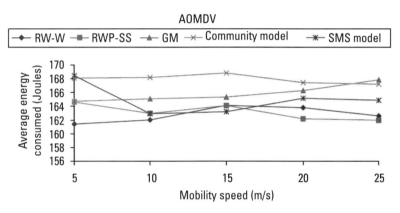

Figure 5.5 Average energy consumed (J) vs. mobility speed (m/s) [31].

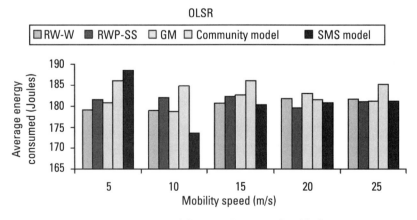

Figure 5.6 Average energy consumed (J) vs. mobility speed (m/s) [31].

model show same quantity of energy consumption. The multipoint relays in OLSR contribute to the energy consumption in OLSR.

The energy consumption of RW-W mobility model is more than RWP-SS and Gauss-Markov mobility models for the FSR routing protocol (Figure 5.7). It is quite unexpected for a random waypoint based mobility model. Energy consumption of FSR under RW-W increases with the increase in mobility speed. The SMS mobility model has the highest energy consumption, alternating between 180 J to 170 J.

The TORA routing protocol has highest energy consumption for the RWP-SS mobility model for up to 10 m/s. After 10 m/s, energy consumption is more for the RW-W mobility model (Figure 5.8). In the literature, it has been shown that the TORA routing protocol consumes more energy. On observing

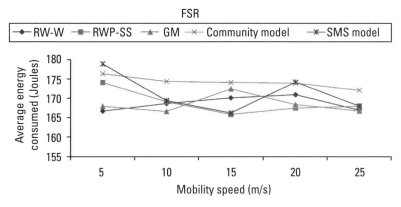

Figure 5.7 Average energy consumed (J) vs. mobility speed (m/s) for FSR routing protocol under different mobility models [31].

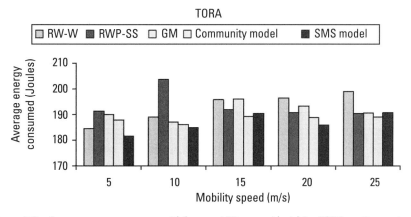

Figure 5.8 Average energy consumed (J) vs. mobility speed (m/s) for TORA routing protocol under different mobility models [31].

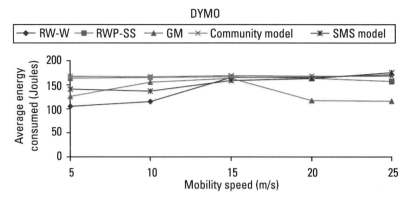

Figure 5.9 Average energy consumed (J) vs. mobility speed (m/s) for DYMO rating protocol under different mobility models [31].

those results, it was found out that those were obtained by having RWP as the underlying mobility model. But the TORA routing protocol performs better under the community mobility model and SMS model. This indicates that the underlying mobility model has the greatest impact on the TORA routing protocol.

For the DYMO routing protocol, both the RWP-SS and community model exhibit the same level of energy consumption (Figure 5.9). The Gauss-Markov mobility model has the least energy consumption. For the DYMO routing protocol under the Gauss-Markov mobility model, the energy consumption decreases with the increase in mobility speed.

References

[1] Timo, R. C., and W. Hanlen, "MANETs: Routing Overhead and Reliability," IEEE, 2006.

[2] Tao, Z., "An Analytical Study on Routing Overhead of Two level Cluster based Routing Protocols for Mobile Ad Hoc Networks," IEEE, 2006.

[3] Broch, J., D. A. Maltz, D. B. Johnson, Y.-C. Hu, and J. Jetcheva, "A Performance Comparison of Multi-Hop Wireless Ad Hoc Network Routing Protocols." In *Proceedings of ACM MobiCom 1998*, October 1998.

[4] Das, S. R., R. Castaneda, and J. Yan, "Simulation-Based Performance Evaluation of Routing Protocols for Mobile Ad Hoc Networks," *Mobile Networks and Applications*, Vol. 5, No. 3, September 2000, pp. 179–189.

[5] Boukerche, A., "Performance Evaluation of Routing Protocols for Ad Hoc Wireless Networks," *Mobile Networks and Applications*, Vol. 9, No. 4, August 2004, pp. 333–342.

[6] Gray, R. S., D. Kotz, C. Newport, N. Dubrovsky, A. Fiske, et al., "Outdoor Experimental Comparison of Four Ad Hoc Routing Algorithms," in *Proceedings of ACM International Symposium on Modeling, Analysis and Simulation of Wireless and Mobile Systems*, October 2004.

[7] Borgia, E., "Experimental Evaluation of Ad Hoc Routing Protocols." In *Proceedings of IEEE International Conference on Pervasive Computing and Communications Workshops*, May 2005.

[8] Cano, J.-C., and P. Manzoni, "A Performance Comparison of Energy Consumption for Mobile Ad Hoc Network Routing Protocols." In *Proceedings of the 8th International Symposim on Modeling, Analysis and Simulation of Computer and Telecommunication Systems*, IEEE Computer Society, 2000.

[9] Feeney, L. M., and M. Nilsson, "Investigating the Energy Consumption of a Wireless Network Interface in an ad Hoc Netwroking Environment." In *Proceedingss of IEEE Conference on Computer Communications (Infocom'01)*, April 2001.

[10] Prabhakaran, P., and R. Sankar, "Impact of Realistic Mobility Models on Wireless Networks Performance," IEEE International Conference on Wireless and Mobile Computing, Networking and Communications, Montreal, June 19–21, 2006, pp. 329–334.

[11] Chen, B.-R., and C. H. Chang, "Mobility Impact on Energy Conservation of Ad Hoc Routing Protocols," SSGRR 2003, Italy, July 28–August 2, 2003.

[12] Kulkarni, S. A., and G R. Rao, "Mobility and Energy Based Performance Analysis of Temporally Ordered Routing Algorithm for Ad Hoc Wireless Network," *IETE Tech Rev*, Vol. 25, 2008, pp. 222–227.

[13] Sucec, J., and I. Marsic, "Hierarchical Routing Overhead in Mobile Ad Hoc Networks," Clustering Overhead For Hierarchical Routing in Mobile Ad Hoc networks," IEEE INFOCOM, 2002.

[14] Qin, Y., and J. He, "The Impact of Throughput of Hierarchical Routing in Ad Hoc Wireless Networks," IEEE, 2005.

[15] Sucec, J., and I. Marsic, "Hierarchical Routing Overhead in Mobile Ad Hoc Networks," *IEEE Transactions on Mobile Computing*, Vol. 3, No. 1, January–March 2004, pp. 46–56.

[16] Mingqiang, I.-I., and W. K. G.Seah, "Analysis of Clustering and Routing Overhead for Clustered Mobile Ad Hoc Networks," IEEE ICDCS, 2006.

[17] Wu, H., and A. Abouzeid, "Cluster-Based Routing Overhead in Networks with Unreliable Nodes." In *Proceedings of IEEE WCNC 2004*, March 2004.

[18] Zhou, N., and A. A. Abouzeid, "Routing in Ad Hoc Networks: A Theoretical Framework with Practical Implications." In *Proceedings of IEEE INFOCOM 2005*, March 2005.

[19] Sprint. "IP Monitoring Project," February 6, 2004, http://ipmon.sprint.com/packstat/packetoverview.php.

[20] Arango, J., S. Ali, and D. Hample, "Header Compression for Ad Hoc Networks," IEEE, 2004.

[21] Wang, H., et al. "A Robust Header Compression Method for Ad Hoc Network," IEEE, 2006.

[22] Rakeshkumar, V., M. Misra, " An Efficient Mechanism for Connecting MANET and Internet Through Complete Adaptive Gateway Discovery," IEEE, 2006.

[23] Ruiz, M., and A. F. G. Skarmet, "Enhanced Internet Connectivity Through Adaptive Gateway Discovery." In *Proc. 29th Annual IEEE International Conference on Local Computer Networks*, November 2004, pp. 370–377.

[24] Chew, K. A., and R. Tafazolli, "Overhead Control in Interworking of MANET and Infrastructure Backed Mobile Networks," International Conference on Wireless Ad Hoc Networks, IEEE, 2004.

[25] Teng, R., H. Morikawa, T. Aoyama, " A Low Overhead Routing Protocol for Ad Hoc Networks with Global Connectivity," IEEE, 2005.

[26] Bansal, S., R. Shorey, and A. Misra, "Comparing the Routing Energy Overheads of Ad-Hoc Routing Protocols," IEEE Wireless Communications and Networking (WCNC 2003), March 20, 2003, New Orleans, LA, 2003.

[27] Rao, V. P., and D. Marandin, "Adaptive Channel Access Mechanism for Zigbee IEEE 802.15.4," *Journal of Communications Software and Systems*, Vol. 2, No. 4, December 2006, pp. 283–293.

[28] Min, R., and A. Chandrakasan, "Energy Efficient Communication for Ad-Hoc Wireless Sensor Networks," *Conference Record of the 35th Asilomar Conference on Signals, Systems and Computers*, 2001, pp 139–143.

[29] Deng, J., et al, "Optimum Transmission Range for Wireless Ad Hoc Networks," 2nd IEEE Upstate New York Workshop on Sensor Networks, Syracuse, NY, October 10, 2002.

[30] International Conference on World Congress on Engineering and Computer Science, 2010, IAENG, USA.

[31] International Journal of Computer Science and Network Security, Vol. 10, No. 5, May 2010.

6

Study of Various Issues in WSNs and Analysis of Wireless Sensor and Actor Network Scenarios

6.1 Introduction

Ad hoc networks are infrastructure-independent networks [1]. So what's a wireless sensor network (WSN)? We have different viewpoints for this question. According to Akylidiz et al. [2], WSN consists of a large number of nodes that are deployed in such a way that they can sense the phenomena. K. Akkaya et al. [3] define WSN as a network that consists of small nodes with sensing, computation, and communication capabilities. We shall generalize these viewpoints and define WSN as a special class of ad hoc wireless networks that are used to provide a wireless communication infrastructure that allows us to instrument, observe, and respond to phenomena in the natural environment and in our physical and cyber infrastructure.

Even though sensor networks are a special type of ad hoc network, the protocols designed for ad hoc networks cannot be used as they are for sensor networks due to the following reasons:

- The number of nodes in sensor networks is very large and has to scale to several orders of magnitude more than the ad hoc networks and thus require different and more scalable solutions.

- The expected data rate is very low in WSN, and the available data is statistical in nature. But mobile ad hoc network (MANET) is designed to carry rich multimedia data and is mainly deployed for distributed computing.
- A sensor network is usually deployed by a single owner, but MANET is usually run by several unrelated entities [4].
- Sensor networks are data centric (i.e., the queries in the sensor network are addressed to nodes that have data satisfying some conditions) and unique addressing is not possible as they do not have global identifiers. But MANET is node centric, with queries addressed to particular nodes specified by their unique addresses.
- Sensor nodes are usually deployed once in their lifetime, and those nodes are generally stationary except for a few mobile nodes, while nodes in MANET move in an ad hoc manner.
- Like MANET, sensor nodes are also designed for self configuration, but the difference in traffic and energy consumption require separate solutions. In comparison to ad hoc networks, sensor nodes have limited power supply and recharge of power is impractical considering the large number of nodes and the environment in which they are deployed. Therefore, energy consumption in WSN is an important metric to be considered.
- Sensor networks are application specific. One can't have a solution that fits for all the problems.
- Simplicity is the rule in the WSN. Since sensor nodes are small and there are restriction on energy consumption, the communicating and computing software in the nodes should be of less size and more computationally efficient than the traditional software used for the same purpose.

In this chapter, we study the various issues associated with WSNs, including software and hardware. A comprehensive study and analysis of various routing protocols in WSAN from an actor-to-actor perspective has been presented. We like to emphasize the routing protocols that we have considered here were designed for ad hoc networks, but we have considered them for wireless sensor and actor networks. This explains the substantial difference between the results that we have obtained by applying the routing protocols designed for ad hoc network to WSAN and the results when the same routing protocols are applied to the ad hoc framework.

6.2 Literature Background

In Parachuri et al. [5], random asynchronous wakeup, an energy management scheme explicitly designed for wireless sensor and actor networks, is introduced. This protocol achieved good scalability while reducing energy consumption.

A novel delay energy–aware routing protocol (DEAP) is presented in Arjan Desai et al. [6] for wireless sensor and actor networks. DEAP provides a flexible range of tradeoffs between the packet delay and the energy use. Therefore, DEAP supports delay-sensitive applications of heterogeneous sensor and actor networks. Han Peng et al. [7] propose the novel localization scheme ECLS: an efficient cooperative localization scheme for wireless sensor and actor networks. ECLS is an event-driven localization method characterized by ideas such as limited beacons and actors cooperation.

A fault-tolerant model is proposed in Keiji Ozake et al. [8] to provide reliable real-time communications among sensors, actors, and actuation devices. The authors incorporate a multiactor/multisensor (MAMS) model. This paper discusses how to make WSAN reliable and available by preventing conflicting actions on multiple actor nodes.

6.3 Applications of WSNs

Sensor networks have been proposed for a variety of applications [9, 10].

- *Intrusion detection and tracking:* The major field where the sensors are finding immediate application is in the military. Sensors are deployed along the border of a battlefield for surveillance, battle damage assessment, nuclear, biological, chemical attack detection, and tracking of intruding personnel and vehicles.
- *Habitat monitoring:* Habitat monitoring includes monitoring climatic changes due to pollution and ice melting, monitoring the nesting habitats of sea birds, and monitoring the microclimate changes in forests due to fires.
- *Motion monitoring:* This includes installing sensors on bridges or large buildings to understand earthquake vibration patterns and providing security by observing the motion of objects in art galleries, shopping malls, or any other facility.
- *Health applications:* Sensors can be used to analyze the physiological conditions of a person and to monitor the drugs administered to patients in the hospital.

- *Traffic analysis:* Traffic sensor networks can monitor and track vehicles on a congested road, as well as detect and monitor car thefts in busy traffic.

More applications are discussed in Chapter 9.

6.4 Hardware and Software Issues in WSN

Wireless sensor networks are composed of hundreds of thousands of tiny devices called *nodes*. A sensor node is often abbreviated as a node. What is a sensor node? A sensor is a device that senses information and passes the same on to a *mote*. Sensors are used to measure the changes to the physical environment like pressure, humidity, sound, and vibration, and changes to the health of a person, like blood pressure, stress, and heartbeat. A mote consists of processor, memory, battery, A/D converter for connecting to a sensor, and a radio transmitter for forming an ad hoc network. A mote and sensor together form a *sensor node*. The structure of the sensor node is shown in Figure 6.1. There can be different sensors for different purposes mounted on a mote. Motes are also sometimes referred to as *smart dust*. A sensor node forms a basic unit of the sensor network [11, 12].

6.4.1 Hardware Issues

The nodes used in sensor networks are small and have significant energy constraints. The hardware design issues of sensor nodes are quite different from other applications and they are as follows [13]:

- Radio range for most of the common radio links allow a mote to transmit at a distance of something like 10 to 200 feet (3 to 61 meters). Radio range is critical for ensuring network connectivity and data collection

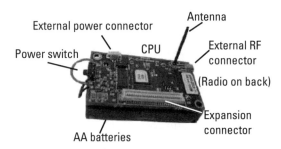

Figure 6.1 The Mica2 sensor node (http://webhosting.devshed.com).

in a network, as the environment being monitored may not have an installed infrastructure for communication. In many networks the nodes may not establish a connection for many days or may go out of range after establishing connection.

- Use of memory chips like flash memory is recommended for sensor networks, as they are nonvolatile, inexpensive, and volatile.

- Energy/power consumption of the sensing device should be minimized, and sensor nodes should be energy efficient since their limited energy resource determines their lifetime. To conserve power, the node should shut off the radio power supply when not in use. Battery type is important since it can affect the design of sensor nodes. A battery protection circuit to avoid overcharge or discharge problem can be added to the sensor nodes.

- Sensor networks consist of hundreds of thousands of nodes. It is preferred only if the node is cheap.

Various platforms are developed considering these design issues. The different platform classes are generic sensing nodes, high-bandwidth nodes, gateway nodes, and special purpose nodes [14]. The sensors that are used for temperature and pressure monitoring are examples of generic sensing nodes (e.g., Mica2, MicaZ, Rene, and Telos). Acoustic, video, and chemical sensors are examples of high-bandwidth nodes requiring intense computation and communication (e.g., BT node, Imotes). Sensor networks are being designed for security purpose, and the sensor nodes of these types of networks contain one or more end points connected to a database to process and store individual sensor readings. These gateway nodes provide an interface into many existing types of networks (e.g., Stargate, Inrysnc Cerfcube, and Pc104 nodes). If a sensor network is designed for security reasons like intrusion detection on the border, then special sensor nodes are used (e.g., spec nodes). Among them is *Berkeley motes*, which is commercially available by Crossbow Technologies, is very popular and used by various research organizations.

Berkeley motes consists of an embedded microcontroller, low-power radio, flash memory, and is powered by two AA batteries. MICA and MICA2 are the most successful families of Berkeley motes. The MICA2 platform is equipped with an Atmel ATmega128L and has a CC1000 transceiver. A 51-pin expansion connector is available to interface sensors. Microcontrollers are used to handle medium access and baseband processing. An event-driven real-time operating system like TinyOS has been implemented to specifically address the concurrency and resource management needs of sensor nodes [15, 16].

MIT is working on the μ-adaptive multidomain power-aware sensors (μAMPS) project [17–19]. The μAMPS sensor node consists of a strong ARM SA1100 processor, a 2.4-GHz transceiver, and a ROM in which a small operating system (μOS), sensor algorithms, and network protocols reside. The μAMPS sensor node includes various power-aware design techniques such as shutdown of inactive components, dynamic voltage, frequency scaling of the processor core, and adjustable radio transmission based on the required range.

Possible research issues are different strategies to improve signal reception and design of low-power, low-cost sensors and processing units. Various schemes to conserve node power consumption, enable node optimization, and simplify modulation schemes may also be considered for sensor nodes. An operating system framework for a sensor node should be able to provide memory management and resource management in a constrained environment.

6.4.2 Software Issues

The various issues in designing an operating system (OS) for sensor networks are as follows:

- In sensor network, a sensor node is mainly responsible for computation of the extracted data from the local environment. It processes the extracted data and manipulates the data as per the requirement of an application. All these activities require real-time response, processing, and routing of the data. So concurrency management is needed in sensor nodes.

- An OS for sensor nodes should be hardware independent and application specific. It should support multihop routing and simple user-level networking abstractions.

- The OS should have built-in features to reduce the consumption of battery energy. Motes cannot be recharged as and when wished due to their small size and low cost requirement, and it should be in a position to enforce limitation on the amount of resources used by each application [19]. The OS should be priority based, and it should give precedence for higher priority events.

- The OS should have an easy programming paradigm. Application developers should be able to concentrate on their application logic instead of being concerned with low-level hardware issues like scheduling, preempting, and networking.

Various operating systems have been designed for sensor nodes, keeping in mind the previously mentioned design issues.

TinyOS is an open source and far by the most popular OS adapted by both the researchers and industry alike for embedded sensor networks. It has been ported on to many platforms and sensor devices. According to the information available on the TinyOS website [20], TinyOS has a component-based architecture enabling rapid innovation and implementation while minimizing the code size as required by the severe memory constraints which is inherent in sensor networks [21]. TinyOS's component library includes network protocols, distributed services, sensor drivers, and data acquisition tools. The execution model of TinyOS supports complex yet safe concurrent operations. TinyOS has been implemented in NesC language [22], which supports the TinyOS component and concurrency model.

The MANTIS operating system [23] was developed by University of Colorado. Here the main aim is to make the task of programming a sensor network closely resemble programming a PC. Mantis OS model has multithreading, preemptive scheduling, and a network stack [24]. Multithreading in MANTIS OS prevents a long-lived task in a sensor node from blocking the execution of the next second time-sensitive task. Hardware is abstracted through a UNIX-like API of device drivers, with a monolithic hardware abstraction. Using MANTIS OS may be easy for a person familiar with programming in a UNIX environment, but it is still uncertain whether the MANTIS OS model is suitable for a resource-constrained environment such as a sensor network.

Nano-Qplus is another operating system proposed for embedded sensor networks [25]. Nano-Qplus is a multithreaded, lightweight, low-power, and dynamic operating system. The Nano-Qplus adapts classical modular and layered design. The key modules found in Nano-Qplus are nano hardware abstraction layer (nHAL), task management, power management, RF message handling, sensing and actuating, time synchronization, and virtual machine. The key design goal of Nano-Qplus is ease of use for a newbie and flexibility for a research expert so that one can adapt the system for her own advanced research needs.

6.5 Issues in Radio Communication

Performance of wireless sensor networks depends on the quality of wireless communication. However, wireless communication in sensor networks is known for its unpredictable nature.

The main design issues for communication in WSNs are as follows:

- Low power consumption in sensor networks is needed to enable a long operating lifetime by facilitating low duty cycle operation and local signal processing.

- Distributed sensing effectively acts against various environmental obstacles, and care should be taken that the signal strength, and consequently the effective radio range, is not reduced by various factors like reflection, scattering, and dispersions.
- Multihop networking may be adapted among sensor nodes to reduce communication link range, and density of sensor nodes should be high.
- Long-range communication is typically point to point and requires high transmission power with the danger of being eavesdropped. So we should consider short-range transmission to minimize the possibility of being eavesdropped.
- Communication systems should include error control subsystems to detect errors and to correct them.

Communication technologies considered for wireless sensor technologies can be classified into single carrier narrowband and spread spectrum technologies [26, 27].

Narrowband communication methods were employed in the earlier sensor systems. Narrowband system operation applies when channel bandwidth, W, is approximately equal to the data rate, R. For a given bandwidth W, the ultimate capacity in an additive white Gaussian noise (AWGN) channel, C, is governed by the Shannon theorem, and can be expressed as

$$C = W \log_2 (1 + S/N)$$

where S and N are the received signal and noise power, respectively. Narrowband systems are more susceptible to interference than broadband systems. Also, regulatory considerations require shared spectrum access, which prevents narrowband technologies from finding applications in some sensor networks. Spread spectrum communication techniques achieve higher signal-to-noise ratio than the narrowband systems but at the cost of excess signal bandwidth. The two most spreading techniques are direct sequence spread spectrum (DSSS) and frequency hopping spread spectrum (FHSS). DSSS technique is adopted in the Zigbee standard, and FHSS technique is adopted in Bluetooth standard.

Research areas include designing low power consuming communication systems and complementary metal oxide semiconductor (CMOS) circuit technique specifically optimized for sensor networks, designing new architectures for integrated wireless sensor systems, and modulation method and data rate selection.

6.6 Design Issues of MAC Protocols

Communication is a major source of energy consumption in WSNs, and MAC protocols directly control the radio of the nodes in the network. MAC protocols should be designed for regulating energy consumption, which in turn influences the lifetime of the node [28]. The various design issues of the MAC protocols suitable for sensor network environment are as follows [29–33]:

- The MAC layer provides fine-grained control of the transceiver and allows on and off switching of the radio. The design of the MAC protocol should have this switching mechanism to decide when and how frequently the on and off mechanism should be done. This helps in conserving energy.
- A MAC protocol should avoid collisions from interfering nodes, overemitting, overhearing, control packet overhead, and idle listening. When a receiver node receives more than one packet at the same time, these packets are called *collided packets*, and they need to be sent again, thereby increasing energy consumption. When a destination node is not ready to receive messages, then it is called overemitting. Overhearing occurs if a node picks up packets that were destined for some other node. Sending and receiving of less useful packets results in increased control overhead. Idle listening is an important factor, as the nodes often hear the channel for possible reception of the data that has not been sent.
- Scalability, adaptability, and decentralization are additional important criteria in designing a MAC protocol. The sensor network should adapt to the changes in the network size, node density, and topology. Also, some nodes may die over time, some may join, and some may move to different locations. A good MAC protocol should accommodate these changes to the network.
- A MAC protocol should have minimum latency and high throughput when the sensor networks are deployed in critical applications.
- A MAC protocol should include message passing. Message passing means dividing a long message into small fragments and transmitting them in a burst. Thus, a node that has more data gets more time to access the medium.
- There should be uniformity in reporting the events by a MAC protocol. Since the nodes are deployed randomly, nodes from highly dense area may face high contention among themselves when reporting events, resulting in high packet loss. Consequently, the sink detects fewer events

from such areas. Also, the nodes that are nearer to the sink transmit more packets at the cost of nodes that are away from the sink.
- The MAC protocols should take care of the well-known problem of information asymmetry, which arises if a node is not aware of packet transmissions two hops away.
- MAC protocols should satisfy the real-time requirements. MAC is the base of the communication stack, and timely detection, processing, and delivery of the information from the deployed environment is an indispensable requirement in a WSN application.

6.6.1 Classification of MAC Protocols

MAC Protocols for WSNs can be broadly classified into four categories [34]: (a) scheduling based, (b) collision free, (c) contention based, and (d) hybrid schemes.

In a scheduling based MAC protocol, a scheduling algorithm determines the time at which a node can transmit a packet, so that multiple nodes can transmit simultaneously without interference on the wireless channel. Here the time is usually divided into slots, and slots are further organized into frames. Within each frame, a node is assigned at least one slot to transmit. A scheduling algorithm usually finds the shortest possible frame so as to achieve high spatial reuse and low packet latency. TDMA protocols are typical scheduled protocols, and they naturally conserve energy as they have a built-in duty cycle and do not suffer from collisions. TDMA protocols are often required to form a cluster. Some (e.g., scheduling based MAC protocol or TDMA-based protocol) are self-organizing medium access control for sensor networks (SMACS), energy efficient MAC protocol for sensor networks (EMAC), low energy adaptive cluster hierarchy (LEACH), and eavesdrop and register (EAR).

Some collision-free MAC protocols like spatial TDMA, implicit prioritized access protocol, and traffic-adaptive medium access protocol have been designed. They are employed in WSNs where frequent transport of traffic occurs.

Contention based MAC protocols are based on carrier sense multiple access with collision avoidance mechanism. Here, nodes are not required to form a cluster. These protocols inherit good scalability and support node changes and new node inclusions. The standardized IEEE 802.11 distributed coordinated function (DCF) is an example of the contention based protocol, and it is robust to the hidden node terminal problem. Some MAC protocols that are designed as contention based are sensor-MAC (S-MAC), Time-MAC (T-MAC), and WiseMAC.

MAC protocols such as the physical layer driven protocol, hybrid TDM-FDM MAC, and channel assignment table based MAC are examples of hybrid schemes.

S-MAC protocol is an effective, energy-conserving MAC protocol for wireless sensor networks. S-MAC accommodates the various design issues of MAC protocols, which we have discussed previously. A majority of the contention based MAC protocols are based on S-MAC. Conserving energy in S-MAC protocol is done through three novel techniques. A low-duty cycle is implemented in S-MAC, which forces the nodes to sleep periodically instead of listening continuously to an idle channel. Transceivers are turned off for the time when the shared medium is used for transmission by other nodes. A message-passing scheme is used for applications that require store-and-forward processing, thereby reducing latency and control overhead.

The basic idea of S-MAC is that time is divided into large frames. Every frame starts off with a small synchronization phase, followed by a fixed active part and sleep part. During synchronization phase, nodes receive or send sync packet containing the schedule information (i.e. when to sleep). During the active part, it can communicate with its neighbors and send any messages queued during the sleep part. During the sleep part, a node turns off its radio to conserve energy.

Instead of using the whole frame, all the messages are packed into the active part, thus reducing the energy wasted due to idle listening. S-MAC also tries to avoid overhearing by letting interfering nodes to go to sleep after they hear an RTS or CTS packet. S-MAC utilized virtual clustering technology to provide good scalability.

Conserving energy at the MAC layer by reducing potential energy waste is a very important research area. Lot of research has to be done for fine tuning the radio parameters, which are the main source of energy consumption.

MAC protocols should be more susceptible to the movement of nodes. The research community generally ignores mobility at the MAC layer because sensor networks were originally assumed to compromise static nodes. But recent works like Robo Mote [35] have enabled mobility in sensor nodes, and there is much room for research in this area. Optimization criteria such as latency, compliance with real-time constraints, or reliable data delivery for MAC protocols have not gained importance in research.

The cross layer designs in sensor networks have lead to monolithic, vertically integrated solutions, which might work for a research group but may not be useful for other research groups. So developing standard sensor network architecture is a continuous process.

6.7 Deployment

Deployment means setting up an operational sensor networks in a real-world environment [36]. Deployment of sensor network is a labor intensive and cumbersome activity, as we do not have influence over the quality of wireless communication, and the real world puts strains on sensor nodes by interfering during communications. Sensor nodes can be deployed either by placing one after another in a sensor field or by dropping them from a plane.

Various deployment issues that need to be taken care are as follows [37, 38]:

- When sensor nodes are deployed in the real world, node death due to energy depletion either caused by normal battery discharge or due to short circuits is a common problem that may lead to inaccurate sensor readings. Also, sink nodes act as gateways, and they store and forward the data collected. Hence, problems affecting sink nodes should be detected to minimize data loss.
- Deployment of sensor networks results in network congestion due to many concurrent transmission attempts made by several sensor nodes. Concurrent transmission attempts occur due to inappropriate design of the MAC layer or by repeated network floods. Another issue is the physical length of a link. Two nodes may be very close to each other, but still they may not be able to communicate due to physical interference in the real world, while nodes that are far away may communicate with each other.
- Low data yield is another common problem in real-world deployment of sensor nodes. Low data yield means a network delivers insufficient amount of information.
- Self configuration of sensor networks without human intervention is needed due to random deployment of sensor nodes.

A framework is proposed in [38] considering these deployment issues. POWER is a software environment for planning and deploying wireless sensor network applications into actual environment. POWER has three main parts: network deployment, simulation and performance evaluation, and optimization. The network deployment aims to find an optimal placement solution. To support WSN deployment solution for a real application, it is needed to simulate the WSN and carry out the quantitative analysis of the WSN. The evaluation part focuses on evaluating the virtual WSN, and optimization works on optimizing the network performance that can be found after evaluating the network performance. The various metrics considered for evaluating the virtual WSN are lifetime, latency, scalability, and reliability.

Research issues include improving the range and visibility of the radio antennas when deployed in various physical phenomena and detecting inaccurate sensor readings at the earliest time possible to reduce latency and reduce congestion.

6.8 Localization

Sensor localization is a fundamental and crucial issue for network management and operation. In many of the real-world scenarios, the sensors are deployed without knowing their positions in advance and there is also no supporting infrastructure available to locate and manage them once they are deployed [39–41].

Determining the physical location of the sensors after they have been deployed is known as the problem of localization. Location discovery or localization algorithm for a sensor network should satisfy the following requirements [42]:

- The localization algorithm should be distributed since a centralized approach requires high computation at selective nodes to estimate the position of nodes in the whole environment. This increases signaling bandwidth and also puts extra load on nodes close to the center node.
- Knowledge of the node location can be used to implement energy-efficient message routing protocols in sensor networks.
- Localization algorithms should be robust enough to localize the failures and loss of nodes. They should be tolerant to error in physical measurements.
- It is shown in [43] that the precision of the localization increases with the number of beacons. A beacon is a node that is aware of its location. But the main problem with increased beacons is that they are more expensive than other sensor nodes, and once the unknown stationary nodes have been localized using beacon nodes then the beacons become useless.
- Techniques that depend on measuring the ranging information from signal strength and time of arrival require specialized hardware that is typically not available on sensor nodes.
- Localization algorithms should be accurate, scalable, and support mobility of nodes.

Most of the localization algorithms have two phases, ranging phase and refinement phase. In ranging phase, sensors do initial estimation of their positions by using some ranging technologies such as received signal strength (RSS), direction of arrival (DoA), time of arrival (ToA), or time difference of arrival (TDoA). Refinement phase helps in refining the estimated position leading toward correct position.

The research on mobile nodes localization and motion analysis in real time will continue to grow as sensor networks are deployed in large numbers and as applications become varied. Scientists in numerous disciplines are interested in methods for tracking the movements and population counts of animals in their habitats (i.e., passive habitat monitoring). Another important application is to design a system to track the location of valuable assets in an indoor environment. We need to improve the maximum likelihood estimation in a distributed environment like sensor networks. Developing mobile assisted localization is another important research area. One needs to improve the localization accuracy, which depends on ToA or TDoA.

6.9 Synchronization

Clock synchronization is an important service in sensor networks. Time synchronization in a sensor network aims to provide a common timescale for local clocks of nodes in the network. A global clock in a sensor system will help process and analyze the data correctly and predict future system behavior. Some applications that require global clock synchronization are environment monitoring, navigation guidance, and vehicle tracking. A clock synchronization service for a sensor network has to meet challenges that are substantially different from those in infrastructure based networks [44–46].

- Energy utilization in some synchronization schemes is more due to energy-hungry equipment like GPS receivers or network time protocol (NTP).
- The lifetime or the duration for the nodes that are spread over a large geographical area needs to be taken into account. Sensor nodes have a higher degree of failure. Thus, the synchronization protocol needs to be more robust to failures and to communication delay.
- Sensor nodes need to coordinate and collaborate to achieve a complex sensing tasks like data fusion. In data fusion, the data collected from different nodes are aggregated into a meaningful result. If the sensor nodes lack synchronization among themselves, then the data estimation will be inaccurate.

- Traditional synchronization protocols try to achieve the highest degree of accuracy. The higher the accuracy, the bigger requirement for resources will be. Therefore, we need to have a tradeoff between synchronization accuracy and resource requirements based on the application.
- Sensor networks span multihops with higher jitter. So, the algorithm for sensor network clock synchronization needs to achieve multihop synchronization even in the presence of high jitter.

There are many different types of clock synchronization:

- *Global clock:* A precise global clock called UTC is maintained by atomic clocks in standard laboratories. Maintaining this time in sensor networks is significantly harder, and sensor networks also typically do not require this strict clock synchronization.
- *Relative notion of clock:* This is the relative notion of time within the sensor network. Each node in a sensor network is synchronized with every other node with a logical notion but not with the real time. Here, the logical notion need not match the physical clock.
- *Physical ordering:* If ordering of events is more important than the precise time (i.e., if the system is able to state whether an event occurred before or after another event), then this type of synchronization involves ordering of events in some partial or total order.

Various synchronization protocols can be found in the literature. The reference broadcast synchronization (RBS) protocol seeks to reduce nondeterministic latency using receiver-to-receiver synchronization. Delay measurement time synchronization protocol for wireless sensor network is an energy efficient protocol due to its low message complexity. The probabilistic clock synchronization service for sensor networks extends RBS by providing probabilistic bounds on the accuracy of clock synchronization.

Various research issues include building an analytical model for multihop synchronization and improving the radio communication in the existing synchronization protocols like RBS and lightweight tree based synchronization (LTS).

6.10 Calibration

Calibration is the process of adjusting the raw sensor readings obtained from the sensors into corrected values by comparing them with some standard values.

Manual calibration of sensors in a sensor network is a time-consuming and difficult task due to failure of sensor nodes and random noise, which makes manual calibration of sensors too expensive.

Various calibration issues in sensor networks are as follows [47–49]:

- A sensor network consists of large number of sensors typically with no calibration interface.
- Access to individual sensors in the field can be limited.
- Reference values might not be readily available.
- Different applications require different calibration.
- Calibration is required in a complex dynamic environment with many observables like aging, decaying, and damage.
- Other objectives of calibration include accuracy, resiliency against random errors, ability to be applied in various scenarios, and ability to address a variety of error models.

Calibration can be of micro and macro variety. In micro-calibration, each device is individually tuned in a carefully controlled environment. This type of calibration can be performed in the factory during the production stage or manually in the field. Micro-calibration can be problematic in sensor networks that consist of thousands of sensor nodes, as it is not possible to calibrate each and every device in the sensor network. This has resulted in macro-calibration for sensor networks. In [49] the authors approach macro-calibration by framing it as a parameter estimation problem. Here, for each device calibration parameters are chosen in such a way that they optimize the overall system response, instead of the individual device response. Some of the benefits of this technique are that it frees the person from observing each and every device for calibration since we only need to observe the overall system and that it provides calibration interface to those devices that do not already have one. Research includes designing various calibration techniques involving the various issues discussed previously.

6.11 Network Layer Issues

Over the past few years sensor networks are being built for specific applications, and routing is important for sending the data from sensor nodes to the base station (BS). The BS plays a very important role in communicating with other sensor nodes in the network and is used for processing the data gathered from the sensor nodes found in the vicinity of the phenomena. As discussed in the

Introduction, routing in sensor networks is a very challenging issue. Various issues at the network layer are [50–52].

Energy efficiency is a very important criterion. We need to discover different techniques to eliminate energy inefficiencies that may shorten the lifetime of the network. At the network layer, we need to find various methods for discovering energy-efficient routes and for relaying the data from the sensor nodes to the BS so that the lifetime of a network can be optimized.

Routing protocols should incorporate multipath design technique. Multipath refers to those protocols that set up multiple paths so that a path among them can be used when the primary path fails.

Path repair is desired in routing protocols whenever a path break is detected. Fault tolerance is another desirable property for routing protocols. Routing protocols should be able to find a new path at the network layer even if some nodes fail or are blocked due to environmental interference.

Sensor networks collect information from the physical environment and are highly data centric. In the network layer in order to maximize energy savings we need to provide a flexible platform for performing routing and data management.

The data traffic that is generated will have significant redundancy among individual sensor nodes since multiple sensors may generate same data within the vicinity of a phenomenon. The routing protocol should exploit such redundancy to improve energy and bandwidth utilization.

As the nodes are scattered randoml,y resulting in an ad hoc routing infrastructure, a routing protocol should have the property of multiple wireless hops.

Routing protocols should take care of the heterogeneous nature of the nodes (i.e., each node will be different in terms of computation, communication, and power).

6.11.1 Classification of Routing Protocols in WSNs

Routing protocols for WSNs in general can be classified based on the network structure [3]: (a) flat based routing, (b) hierarchical based routing, and (c) location based routing.

If all the nodes are assigned equal roles or functionality, then it is flat based routing (e.g., sensor protocols for information via negotiation, or SPIN, and rumor routing, direct diffusion).

If the nodes play different roles in the network, then it is hierarchical based routing (e.g., LEACH and TEEN).

If a sensor node's position is exploited to route data in the network, then it is location based routing (e.g., geographic and energy aware routing, or GEAR, and sequential assignment routing, or SAR).

Furthermore, routing protocols can be further classified based on network operation:

- Multipath based;
- Query based;
- Negotiation based;
- QoS based routing.

Multipath routing protocols use multiple paths instead of a single path to enhance network performance (e.g., directed diffusion).

In query based routing a destination node propagates a query for some data throughout the network. Then a node that has data that matches the query sends back the data to the node that had initiated the query. Usually, these queries are described in some high-level language or natural language (e.g., rumor routing).

Negotiation based routing protocols use high-level data descriptors in order to eliminate redundant data transmission through negotiation (e.g., SPIN).

In QoS based routing, the routing protocols need to satisfy some QoS metrics, such as delay, energy, or bandwidth, when delivering the data to base station (e.g., SAR).

If certain parameters in routing protocols can be controlled in order to make it adapt to the current network conditions and energy levels, then such routing protocols are called *adaptive routing protocols*.

Routing protocols are classified, depending on how the source node finds the route to the destination node, as (a) reactive, (b) proactive, and (c) hybrid routing protocols.

In proactive routing protocols all the routes are computed before they are actually needed, while a reactive routing protocol is an on-demand routing protocol. A combination of proactive and reactive routing protocols results in hybrid routing protocols.

6.11.2 Popular Routing Protocols

Some popular routing protocols are explained as follows:

- *Low energy adaptive cluster hierarchy:* LEACH is proposed in [53]. This WSN is considered to be a dynamic clustering method network and is made up of nodes, some of which are called *cluster heads*. The job of the cluster head is to collect data from its surrounding nodes and pass it on to the base station. LEACH is dynamic because the job of cluster head rotates. The reason we need this network protocol is due to the fact that a node in the network is no longer useful when its battery dies, so this allows us to space out the lifespan of the nodes, allowing it to do only the minimum work it needs to transmit data. Cluster heads are chosen randomly to employ the technique of randomly rotating the role of a cluster head among all the nodes in the network. The operation of LEACH is organized in two phases, where each phase consists of a setup and a transmission. During the setup phase, the nodes organize themselves into clusters with one node serving as the cluster head in each cluster. The decision to become a cluster head is made locally within each node, and a predetermined percentage of the nodes serve as local cluster heads in each round on average. During the transmission phase, the self-elected cluster heads collect data from nodes within their respective clusters and apply data fusion before forwarding them directly to the base station. At the end of a given round, a new set of nodes becomes cluster heads for the subsequent round. Furthermore, the duration of the transmission phase is set much larger than that of the setup phase in order to offset the overhead due to cluster formation. Thus, LEACH provides a good model where localized algorithms and data aggregation can be performed within randomly self-elected cluster heads, which helps in reducing information overload and providing a reliable set of data to the end user. It has been shown that LEACH provides significant energy savings and a prolonged network lifetime.

- *Directed diffusion:* Directed diffusion, proposed in [54], is a data-centric communication protocol for wireless sensor scenarios. The data generated by the producer is named using attribute-value pairs. The consumer node requests the data by periodically broadcasting an interest for the named data, which in turn depends on the attributes of the data. Each node in the network will establish a gradient toward its neighboring nodes from which it receives the interest. Interest is a descriptor of some event in which the node is interested. The gradient specifies both the data rate and the direction toward which the data should be sent. Once the producer detects an interest, it will send exploratory packets toward the consumer, possibly along multiple paths. As soon as the consumer begins receiving exploratory packets from the producer it will reinforce the path to a particular neighbor from whom it chooses to receive the rest of the data. The data will then flow back toward the consumer along the reinforced path.

In [55] results are obtained by applying ad hoc routing protocols for wireless sensor and actor networks. Here we would like to point out the huge difference between our work and the mentioned work. They have considered packet delivery and delay as the metrics, while we have considered packet delivery, end-to-end delay, overhead, and throughput as the metrics. In their work the authors have mapped the results against simulation time, while we have mapped the results against a number of actor nodes and simulation time. In the mentioned work the simulation time is varied, starting from 30 to 100 sec in steps of 10 while we have varied the simulation time from 20 to 100 sec in steps of 20. They have only two plotted resultant graphs, while we have eight plotted resultant graphs. This shows that we have improved upon their work and have done a detailed comprehensive analysis compared to the mentioned paper.

Sensor networks are still at an early stage in terms of technology, as they are still not widely deployed in the real world, and this opens many doors for research. The current routing protocols need to be improved, as they have their own set of problems.

6.12 Transport Layer Issues

End-to-end reliable communication is provided at transport layer. The various design issues for transport layer protocols are as follows [56, 57]:

- In the transport layer the messages are fragmented into several segments at the transmitter and reassembled at the receiver. Therefore, a transport protocol should ensure orderly transmission of the fragmented segments.
- Limited bandwidth results in congestion, which impacts normal data exchange and may also lead to packet loss.
- Bit error rate also results in packet loss and also wastes energy. A transport protocol should be reliable for delivering data to a potentially large group of sensors under extreme conditions.
- End-to-end communication may suffer due to various reasons, as the placement of nodes is not predetermined and external obstacles may cause poor communication performance between two nodes. If this type of problem is encountered, then end-to-end communication will suffer. Another problem is failure of nodes due to battery depletion.
- In sensor networks the loss of data, when it flows from source to sink, is generally tolerable. But the data that flows from sink to source is sensitive to message loss. (A sensor obtains information from the surrounding environment and passes it on to the sink, which in turn queries the sensor node for information.)

Traditional transport protocols such as UDP and TCP cannot be directly implemented in sensor networks for the following reasons:

- If a sensor node is far away from the sink, then the flow and congestion control mechanism cannot be applied for those nodes.
- Successful end-to-end transmissions of packets are guaranteed in TCP but it's not necessary in an event-driven applications of sensor networks.

Overhead in a TCP connection does not work well for an event-driven application of sensor networks. UDP, on the other hand, has a reputation of not providing reliable data delivery and has no congestion or flow control mechanisms, which are needed for sensor networks.

The pump slowly, fetch quickly (PSFQ) proposal [58] is based on slowly injecting packets into the network but performing aggressive hop by hop recovery in case of packet losses. PSFQ is a scalable and robust protocol. It is scalable because it supports minimum signaling and is robust because it is responsive to a wide range of operational error conditions found in a sensor network, allowing it to function successfully in an error-prone and highly demanding environment like a sensor network. PSFQ has a hop-by-hop error recovery implementation in which intermediate nodes also take responsibility for loss detection and recovery, so reliable data exchange is done on a hop-by-hop basis rather than end to end. Thus, the hop-by-hop approach is more tolerable to errors. The pump operation is important in controlling the timely dissemination of code segments to all the destination nodes and for providing basic flow control mechanisms so that the retasking operation does not utilize all the resources of the sensor network. A node goes into a fetch mode in PSFQ protocol once a sequence number gap in a file's fragments is detected. A fetch operation requests retransmission from the neighboring nodes once loss is detected at a receiving node.

Developing transport protocols for sensor networks is itself a difficult task due to the previously discussed issues, and not much work is reported.

Existing transport layer protocols for WSNs assume that the network layer uses a single path routing, and multipath routing is not considered. This opens many doors for research in this direction. Many of the transport protocols do not consider priority when routing. Since sensor nodes are placed in various types of environments, the data from different locations will have different priorities.

6.13 Data Aggregation and Dissemination

Data gathering is the main objective of sensor nodes. The sensors periodically sense the data from the surrounding environment, process it, and transmit it to the base station or sink. The frequency of reporting the data and the number of sensors that report the data depends on the particular application. Data gathering involves systematically collecting the sensed data from multiple sensors and transmitting the data to the base station for further processing. But the data generated from sensors is often redundant; also, the amount of data generated may be too huge for the base station to process. Hence we need a method for combining the sensed data into high-quality information, and this is accomplished through data aggregation [59]. Data aggregation is defined as the process of aggregating the data from multiple sensors to eliminate redundant transmission and estimating the desired answer about the sensed environment, then providing fused information to the base station.

Data aggregation protocols based on network architecture can be classified as follows [59–61]:

- Flat networks:
  - Push diffusion;
  - Two phase pull diffusion;
  - One phase pull diffusion;
- Hierarchical networks:
  - Cluster based data aggregation;
  - Chain based data aggregation;
  - Tree based data aggregation;
  - Grid based data aggregation.

Data aggregation is accomplished in flat networks by data-centric routing. In the push diffusion scheme, source nodes initiate the diffusion while the base station responds to the source. Sensor protocol for information via negotiation (SPIN) is an example of a push diffusion scheme. Directed diffusion protocol represents two phase pull diffusion. Directed diffusion is data centric in that all communication is for named data. Directed diffusion is appropriate for applications with many sources and few sinks. One major drawback of two phase pull diffusion is large overhead. To overcome the disadvantages of two phase pull diffusion, the one phase pull diffusion was proposed. Here, sources do not transmit exploratory data resulting in less overhead.

In view of scalability and energy efficiency, the hierarchical data-aggregation techniques are proposed. In cluster based data aggregation, each sensor in the cluster transmits data to its cluster head, which aggregates data from all the sensors and transmits the aggregated data to the sink. Some protocols that come under this scheme are LEACH and HEED. One major drawback of cluster based data aggregation is that if the sensor is far away from the cluster head, then it might spend more energy for communicating with the cluster head. To overcome this drawback, chain based data aggregation protocols were proposed. In chain based data aggregation, data is transmitted only to its nearby neighbor. Power-efficient data gathering protocol for sensor information systems (PEGASIS) is one of the protocols under this scheme. In tree based data aggregation, sensor nodes are organized into a tree. Aggregation is performed at intermediate nodes, and a concise data is sent to the root node. One important application of tree based data aggregation is in monitoring the radiation level in a nuclear plant where a maximum value provides the most useful information for the safety of the plant. In grid based data aggregation, a region is monitored by dividing it into several grids and sensors are placed each grid.

Some design issues in data aggregation are as follows [62]:

- Sensor networks are inherently unreliable, and certain information may be unavailable or expensive to obtain (e.g., the number of nodes present in the network, the number of nodes that are responding). Also, it is difficult to obtain complete and up-to date information of the neighboring sensor nodes to gather information.
- What is the best way to make some of the nodes transmit the data directly to the base station or have less transmission of data to the base station to reduce energy?
- What is the best way to eliminate transmission of redundant data using metadata negotiations, as in SPIN protocol?
- What is the best way to improve clustering techniques for data aggregation to conserve energy of the sensors?
- What is the best way to improve in-network aggregation techniques for increased energy efficiency? In-network aggregation means sending partially aggregated values rather than raw values, thereby reducing power consumption.

Based on data delivery, a sensor network can be classified as continuous, event-driven, observer-initiated, and hybrid [63]. In the continuous model, the data is communicated among sensors continuously at a prespecified rate. In the event-driven data model, the sensors report information only if an event of interest occurs. In the observer-initiated model, the sensors report their results in response to an explicit request from the observer. If all the three approaches coexist in the same network, then that model is referred to as hybrid model.

Data dissemination is a process by which data and the queries for the data are routed in the sensor network [64]. Data dissemination is a two-step process. In the first step, if a node is interested in some events, like temperature or humidity, then it broadcasts its interests first to its neighbors periodically and then through the whole sensor network. In the second step, the nodes that have the requested data will send the data back to the source node after receiving the request. The main difference between data aggregation and data dissemination is that in data dissemination all the nodes including the base station can request the data, while in data aggregation all the aggregated data is periodically transmitted to the base station. In addition, data aggregation data can be transmitted periodically, while in data dissemination data is always transmitted on demand.

Flooding is one important protocol that includes data dissemination approach. In flooding, each sensor node that receives a packet broadcasts it to its neighbors, assuming that node itself is not a destination node of the packet and the maximum hop count is not reached. This ensures that the data and queries for data are sent all over the network. In flooding, duplicate messages can be sent to the same node, which is called *implosion*. This occurs when a node receives the same message again and again from several neighbors. In addition, the same event may be sensed by several nodes leading to several neighboring nodes receiving duplicate reports of the same event; this situation is called *overlap*. Finally, many redundant transmissions occur while using flooding, and flooding does not take into account the available energy at the sensor nodes. Flooding results in wasting a lot of a network's resources and decreases the lifetime of the network significantly.

The main research focus in data aggregation is geared toward conserving energy. Other research issues include improving security in data transmission and aggregation, handling tradeoffs in data aggregation (i.e., tradeoffs between different objectives such as energy consumption, latency, and data accuracy), improving quality of service of the data aggregation protocols in terms of bandwidth, and end-to-end delay.

6.14 Database Centric and Querying

Wireless sensor networks have the potential to span and monitor a large geographical area producing massive amount of data. So sensor networks should be able to accept the queries for data and respond with the results.

The data flow in a sensor database is very different from the data flow of the traditional database due to the following design issues and requirements of a sensor network [65–68]:

- The nodes are volatile, since the nodes may get depleted and links between various nodes may go down at any point in time, but data collection should be interrupted as little as possible.
- Sensor data is exposed to more errors than in a traditional database due to interference of signals and device noise.
- Sensor networks produce data continuously in real time and on a large scale from the sensed phenomenon resulting in the need for updating the data frequently, whereas a traditional database is mostly static and centralized in nature.
- Limited storage and scarce energy is another important constraint that needs to be taken care of in a sensor network database, and a traditional database usually consists of plenty of resources and disk space is not an issue.
- The low-level communication primitives in the sensor networks are designed in terms of named data rather than the node identifiers used in the traditional networks.
- The sensor network database can be realized using traditional approaches like the central warehousing approach and distributed approach.

In the central warehousing approach, the first step is to extract the data from the sensor network in a predefined way. It is stored in a database located on a unique front-end server. Subsequently, query processing takes place on the centralized database so the queries are processed without additional communication cost. Some disadvantages of this method are link failure, early depletion of energy, and unnecessary delays.

A distributed database can be energy efficient when the query rate is less than the rate at which data is generated. But traditional distributed database design is unsuitable for large-scale sensor networks because distributed database design maintains a global metadata about data distribution and network topology.

The authors of [69] suggest a fundamentally different on which a sensor database should be modeled. The first approach is to have an in-network implementation of primitive database query operators such as grouping, aggregation, and joins. Here, *in-network* means implementation of each operator in an application-independent way by routing protocols along with some intermediate nodes. Implementing in-network approach requires novel routing mechanisms that consider network resource constraints along with the order in which the database operators are processed. With in-network storage, the data is stored within the network.

The second approach is that instead of having strict semantics associated with the traditional data models and query languages, using a relaxation in the semantics of the database queries to allow approximate results. The sensor network data is a continuous stream of data with some degree of errors, so it is convenient to present approximate semantics.

In sensor networks the contents of data are generally more important than the address of the sensor that has generated the data. This naturally leads to data-centric storage (DCS), in which data is named and the node in which the data is stored is determined by the name associated with them. There is a way of implementing data-centric storage by using geographic hash tables (GHT) [70]. GHT hashes keys into geographic coordinates and stores a key-value pair at a sensor node that is geographically nearest to the hash of its key. The system replicates stored data locally to ensure persistence when nodes fail. Each data has a unique name that is hashed uniformly as a coordinate in the sensing area; it is represented as a two-dimensional plane. GHT implements two operations: *put*, which stores data, and *get*, which retrieves it. In the put operation, the name of data to be stored is first hashed into a location (x, y) in the sensing field. Then, GHT selects the closest sensor to (x, y), which becomes the home node for that data. The home node is selected using greedy-face-greedy (GFG) routing protocol [71]. Each packet starts in the greedy mode, in which it is routed progressively closer to its destination at each loop. When a packet reaches a node s_i whose neighbors are all farther than s_i to the destination, GFG switches to the perimeter mode and the packet is forwarded. As soon as the packet reaches a node closer to the destination than the previous one, it returns to the greedy mode and reaches closer to the destination node. Data retrieval uses a get operation. The name is first hashed into the destination (x, y), then GFG is used to route the request to (x, y). When the request reaches a node in the perimeter of (x, y), the data is returned to the sender.

So what is querying? Querying is a declarative statement requesting a subset of data obtained by computing statistics about that data.

Based on level of aggregation, database queries can be classified into fully aggregated query, unaggregated query, and partially aggregated query [72]. If a query is processed in-network by transmitting a single value from each sensor node on the routing tree, then it is a fully aggregated query. If an intermediate node transmits all the incoming values toward the base station, then it is referred to as an unaggregated query. In a partially aggregated query, the amount of data each node needs to transmit has an upper bound (e.g., a histogram over the data values).

Sensor query template may have the following semantics:

- SELECT: {attributes, aggregates};
- FROM: {Sensordata S};

- WHERE: {predicate};
- GROUP BY: {attributes};
- HAVING: {predicate};
- DURATION: time interval;
- EVERY: time span e.

The SELECT clause specifies attributes and aggregates from sensor records, the FROM clause specifies the distributed relation of sensor type, the WHERE clause filters sensor records by a predicate, the GROUP BY clause gathers sensor records into different partitions according to some attributes [73–75].

Sensor network databases like TinyDBm [76, 77] and Cougar [78] are the dominant architectures to extract and manage data in sensor network.

TinyDB is a query-processing systems for extracting information from a network of sensors. TinyDB runs on the Berkeley mote platform, on top of the TinyOS operating system. In TinyDB there is no need to write embedded C code for sensors, as TinyDB itself provides a simple, SQL-like interface to specify the data you want to extract. When a query is given to TinyDB, it collects the data from the sensing sensors, aggregates the data, and routes toward the sink [76].

TinyDB advocates acquisitional query processing (ACQP), a new query processing technique to support queries that are unique to a sensor network. As in SQL, the queries in TinyDB consist of SELECT, FROM, WHERE, GROUPBY, HAVING blocks to support selection, join, projection, aggregation, and grouping. TinyDB also supports grouped aggregation queries, thus reducing the quantity of data that needs to be transported across the network.

Following is a list of different queries supported by TinyDB:

- *Lifetime-based queries:* Here users may request a specific lifetime query via a QUERY LIFETIME <x> clause, where <x> is a duration in days, weeks, or months.
- *Monitoring queries:* Queries are used to request the value of one or more attributes continuously and periodically (e.g., monitoring the temperature in a house every half an hour).
- *Network health queries:* Queries to check the status of the network itself. This query is particularly important in sensor networks due to their dynamic and volatile nature.

- *Exploratory queries:* These are one-shot queries examining the status of a particular node or set of nodes at a point in time. Along with a SAMPLE PERIOD clause, users may specify the keyword ONCE.
- *Nested queries:* Both events and materialization points provide a form of nested queries. But TinyDB does not support traditional SQL-style nested queries due to the continuous streaming of data.
- *Actuation queries:* These are queries where users need to take some physical action in response to a query. OUPUT ACTION clause is used for this purpose.

Some research areas in sensor database include providing spatio-temporal querying, multiquery optimization, storage placement, designing a distributed long-term networked data storage, invoking low energy communication overhead, various ways of representing the sensor data, processing and distributing query fragments, dealing with communication failures, and designing various models for deploying and managing a sensor database systems.

6.15 Programming Models for Sensor Networks

Currently, programmers are too concerned with low-level details like sensing and node-to-node communication, raising a need for programming abstractions. There is considerable activity for designing programming models for sensor networks due to following issues [76]:

- Since the data collected from the surrounding phenomenon is not for general purpose computing, we need a reactive, event-driven programming model.
- Resources in a sensor network are very scarce, where even a typical embedded OS consuming hundreds of kilobytes is considered too much. Thus, programming models should help programmers in writing energy-efficient applications.
- We need to reduce the runtime errors and complexity since the applications in a sensor network need to run for a long duration without human intervention.
- Programming models should help programmers to write bandwidth-efficient programs and should be accompanied by runtime mechanisms that achieve bandwidth efficiency whenever possible.

Sensor network programming models can be classified into low-level programming models and high-level programming models. Low-level programming models focus on hardware abstraction and allow flexible control of nodes. TinyOS with Nesc and TinyGALS [80] are examples of this category. If low-level programming models are platform centric, then high-level programming models are application centric. High-level programming models are concerned about how easily application logics can be programmed by facilitating collaboration among sensors. High-level programming models can be further classified into group-level abstraction and network-level abstraction or macropragramming. Group-level abstractions provide a set of programming primitives to handle a group of nodes as a single entity. In macroprogramming, instead of focusing on each individual node independently, the network is programmed as one unit. In [81] a state based programming model that employs the macroprogramming approach is discussed. Improving programming ease in languages such as Nesc and galsC [82] provides tremendous opportunities for research.

6.16 Middleware

Middleware for wireless sensor networks should facilitate development, maintenance, deployment, and execution of sensing based applications. WSN middleware can be considered a software infrastructure that glues together the network hardware, operating systems, network stacks, and applications [83]. Various issues in designing middleware for wireless sensor networks are as follows [84–89]:

- Middleware should provide an interface to the various types of hardware and networks supported by primitive operating system abstractions. Middleware should provide new programming paradigms to provide application-specific APIs rather than dealing with low-level specifications.
- Efficient middleware solutions should hide the complexity involved in configuring individual nodes based on their capabilities and hardware architecture.
- Middleware should include mechanisms to provide real-time services by dynamically adapting to the changes in the environment and providing consistent data. Middleware should be adaptable to the devices being programmed, depending on the hardware capabilities and application needs.
- There should be transparency in the middleware design. Middleware is designed for providing a general framework whereas sensor networks

are themselves designed to be application specific. Therefore, we need to have some tradeoff between generality and specificity.

- Sensor network middleware should support mobility, scalability, and dynamic network organization. Middleware design should incorporate real-time priorities. Priority of a message should be assigned at runtime by the middleware and should be based on the context.
- Middleware should support quality of service considering many constraints that are unique to sensor networks like energy, data, mobility, and aggregation.
- Security has become of paramount importance with sensor networks being deployed in mission critical areas like military, aviation, and the medical field.

Several middleware systems have been designed to deal with the aforementioned issues. Mate [90] is a middleware architecture for constructing application-specific virtual machines that execute on top of TinyOS. Using this architecture, developers can easily change instruction sets, execution events, and virtual machine (VM) subsystems. Mate provides a simple programming interface to sensor nodes.

Middleware solutions can be categorized into a database approach and an event based approach.

In the database approach the entire network is viewed as a distributed database where requests are in the form of SQL queries. TinyDB and Cougar are some examples of database approaches in sensor networks.

In an event based approach, the network is a reactive system, which responds to changing events like network messages and timeouts. Sensorware [91] uses this approach.

The design and implementation of a middleware layer for fully realizing the potential of wireless sensor networks is an area that still needs to be investigated further. One needs to design developer-friendly middleware architecture that is not only generic, but also should take care of all the underlying hardware intricacies while helping to reduce the energy consumption and provide adequate quality of support.

6.17 Quality of Service

Quality of service (QoS) is the level of service provided by the sensor networks to its users. The authors of [92] define QoS for sensor networks as the optimum number of sensors sending information toward information-collecting sinks or a base station. Since sensor networks are getting implemented in a greater number of applications, including mission-critical applications such as military applications and nuclear plant monitoring applications, QoS is being given considerable review as the events occurring in these situations are of utmost importance.

The QoS routing algorithms for wired networks cannot be directly applied to wireless sensor networks due to the following reasons:

- The performance of most wired routing algorithms relies on the availability of the precise state information, while the dynamic nature of sensor networks make availability of precise state information next to impossible.
- Nodes in the sensor network may join, leave, and rejoin, and links may be broken at any time. Hence maintaining and re-establishing the paths dynamically, which is a problem in WSN, is not a big issue in wired networks.

Various QoS issues in sensor networks are as follows [93–97]:

- The QoS in WSN is difficult because the network topology may change constantly and the available state information for routing is inherently imprecise.
- Sensor networks need to be supplied with the required amount of bandwidth so that they are able to achieve a minimal required QoS.
- Traffic is unbalanced in sensor networks since the data is aggregated from many nodes to a sink node. QoS mechanisms should be designed for an unbalanced QoS-constrained traffic.
- Many a time, routing in sensor networks needs to sacrifice energy efficiency to meet delivery requirements. Even though multihops reduce the amount of energy consumed for data collection, the overhead associated with it may slow down the packet delivery. Also, redundant data makes routing a complex task for data aggregation affecting thus affecting QoS in WSN.
- Buffering in routing is advantageous, as it helps to receive many packets before forwarding them. But multihop routing requires buffering of huge amount of data. This limitation in buffer size will increase the delay variation that packets incur while traveling on different routes and even on the same route, making it difficult to meet QoS requirements.
- QoS designed for WSN should be able to support scalability. Adding or removing of the nodes should not affect the QoS of the WSN.

One of the very first protocols that had QoS support is the sequential assignment routing (SAR) [98]. The area of sensor network QoS largely remains a partially unexplored area. Designing an appropriate QoS model, deciding how many layers need to be integrated, determining support for heterogeneous nodes, designing QoS models for specific applications, designing QoS based protocols in which to integrate with other network like cellular, LANs, and IPs, and designing QoS via middleware layer are prime areas to be investigated and explored in future.

6.18 Security

Security in sensor networks is as much an important factor as performance and low-energy consumption in many applications. Security in a sensor network is very challenging, as WSN is not only being deployed in battlefield applications but also for surveillance, building monitoring, burglar alarms, and in critical systems such as airports and hospitals.

Since sensor networks are still a developing technology, researchers and developers agree that their efforts should be concentrated in developing and integrating security from the initial phases of sensor applications development; by doing so, they hope to provide a stronger and more complete protection against illegal activities and maintain stability of the systems at the same time.

Following are the basic security requirements to which every WSN application should adhere [99–105]:

- Confidentiality is needed to ensure sensitive information is well protected and not revealed to unauthorized third parties. Confidentiality is required in sensor networks to protect information traveling between the sensor nodes of the network or between the sensors and the base station; otherwise, it may result in eavesdropping on the communication.
- Authentication techniques verify the identities of the participants in a communication. In sensor networks, it is essential for each sensor node and the base station to have the ability to verify that the data received was really sent by a trusted sender and not by an adversary that tricked legitimate nodes into accepting false data. A false data can change the way a network could be predicted.
- Lack of integrity may result in inaccurate information. Many sensor applications, such as pollution and healthcare monitoring, rely on the integrity of the information to function (e.g., it is unacceptable to have improper information regarding the magnitude of the pollution that has occurred).

- One of the many attacks launched against sensor networks is the message reply attack, where an adversary may capture messages exchanged between nodes and reply to them later to cause confusion to the network. So sensor networks should be designed for freshness, meaning that the packets are not reused, thus preventing potential mix-up.
- In sensor networks, secure management is needed at the base station level, since all communication in sensor networks ends up at the base station. Issues like key distribution to sensor nodes in order to establish encryption and routing information need secure management. Also, clustering techniques require secure management as well, since each group of nodes may include a large number of nodes that need to be authenticated with each other and exchange data in a secure manner. Security and QoS are two opposite poles in sensor networks. Security mechanisms like encryption should be lightweight so that the overhead is minimized and should not affect the performance of the network.

The security issues posed by sensor networks are a rich field for research problems, including designing routing protocols with built-in security features, a new symmetric key cryptography for sensor networks, secure data aggregation protocols, intrusion detection systems, and security systems for multimedia sensors.

Different types of threats in sensor networks and their brief descriptions are given in Table 6.1.

6.19 Wireless Sensor and Actor Network Architecture

The next step of evolution from wireless sensor network is wireless sensor and actor network (WSAN). WSAN is intertwined by both sensor and actor nodes; where sensor nodes are low-powered nodes with less communication capabilities, actor nodes are technically superior to sensor nodes with high-energy batteries and long-range communication capabilities. In WSN the coordination is between the various sensor nodes and the sink is shown in Figure 6.2. The functionality of the sink is to collect and process the reported data.

As shown in Figure 6.3, two types of WSAN coordination take place: sensor-to-actor and actor-to-actor coordination. Once the actor nodes receive raw information from the sensor nodes about the detected phenomenon, then the actor nodes should process the data and take the required action appropriately. For example, motion monitoring is done to provide security by observing the motion of objects in art galleries, shopping malls, museums, or any other facility. If the sensor nodes detect any movement of artifact in the museum, then

Table 6.1
Attacks in Wireless Sensor Networks

Types of Attacks	Description
Passive information gathering	An intruder with an appropriate receiver and well-designed antenna can easily pick off the data stream and intercept the messages containing the physical locations of the nodes, allowing the attacker to locate and destroy the nodes.
Node subversion	A particular node might be captured and information stored on it might be obtained by an adversary.
Spoofing and altering the routing information	The adversary targets the routing information that is supposed to be exchanged between the nodes. By spoofing, altering, or replaying routing information, adversaries may be successful in creating routing loops and generating false error messages.
Sinkhole attacks	Here the adversary's goal is to lure nearly all the traffic from a particular area through a compromised node, creating a sinkhole with the adversary at the center.
Sybil attacks	A single node presents multiple identities to other nodes in the network.
Wormholes	An adversary tunnels the messages received in one part of the network over a low latency link and replays them in different part of the network.
Hello flood attacks	Many protocols require nodes to broadcast hello packets to announce themselves to their neighbors, and a node receiving such a packet may assume that it is within normal radio range of the sender. But this assumption may be false.
Denial of service attacks	A denial of service attack is any event that diminishes or eliminates a network's capacity to perform its expected function. Hardware failures, software bugs, and resource exhaustion can cause denial of service.
Selective forwarding	A malicious node may refuse to forward certain messages and simply drop them, ensuring that they are not further propagated.
Jamming	In this type of attack, adversaries interfere with the communication frequencies of the sensor nodes present in the network.

the same event is conveyed to the actor nodes, which in turn take appropriate steps like sounding the burglary alarm or informing the police of the burglary. Routing protocols like AODV, DSDV, and DSR designed for ad hoc networks can be applied for communication between actor-to-actor nodes of WSAN as long as the real-time requirements are met and the communication overhead occurring at the sensor nodes due to actor-to-actor communication is kept low [106, 107].

6.20 Simulation Environment

The simulation code has been modified wherever it was deemed necessary to satisfy our simulation conditions. Also several parameters have been fine tuned as specified by Stuart et al. [108]. For our simulation, we have used NS2 and NRL Sensorsim [109] in combination. NRL Sensorsism was developed at Naval Research Laboratory (NRL) to extend the NS2 capability to simulate sensor

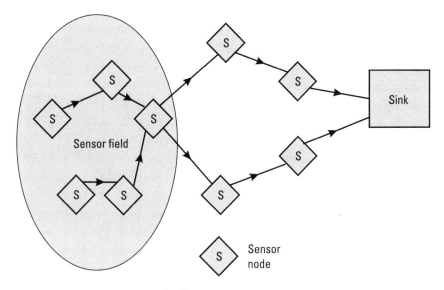

Figure 6.2 Sensor nodes to sink [107].

networks. The simulated sensor area is a 501m × 501m rectangle. In the first run, the number of nodes has been varied from 50 to 100 nodes, keeping the simulation time constant at 100 sec. In the second run, the number of nodes has been kept constant at 100 nodes, while varying the simulation time from 20 to 100 sec. The MAC layer protocol is a modified IEEE 802.11, which confirms to a sensor network environment. The size of each of the message transmitted is 100 bytes. The transmission range is 50m.

Various performance metrics like PDR, average end-to-end delay, protocol control overhead, and throughput of the network have been selected to evaluate the performance of the routing protocols.

6.21 Result Analysis

The network performance is studied under simulation environments for various routing algorithms considering various metrics like PDR, average end-to-end delay (AEED), protocol control overhead (PCO), and throughput of the network.

6.21.1 Packet Delivery Ratio Based Performance Evaluation

The variation of PDR with number of actor nodes and simulation times are depicted in Figures 6.4 and 6.5, respectively. All three protocols show a decreasing tendency for PDR values with the increasing number of nodes.

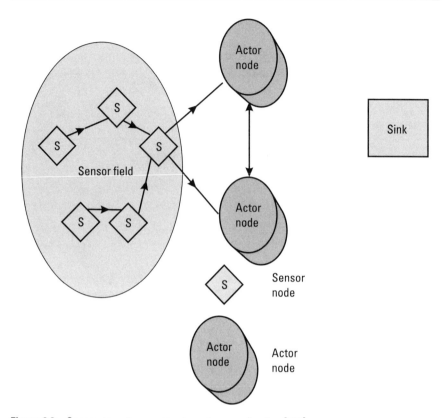

Figure 6.3 Sensor-to-actor or actor-to-actor coordination [107].

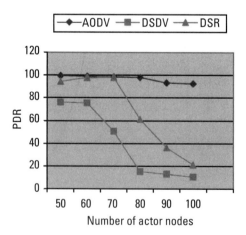

Figure 6.4 PDR vs. number of actor nodes [107].

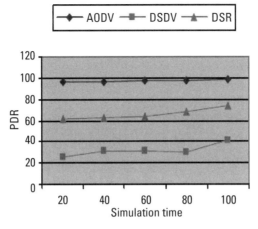

Figure 6.5 PDR vs. simulation time.

Initial PDR of DSR can be compared to that of AODV, but there is a freefall in PDR once the number of nodes increases to greater than 70, which can be attributed to the number of dropped packets and the amount of collisions that occur in the network. Still, DSR is better than DSDV. As can be seen in the graph, AODV has the highest PDR when compared to other protocols, while DSDV is the worst performer. We find a little fall in PDR of AODV as the number of actor nodes is increased, since a packet sent from one actor node to actor node will have more hops to traverse before reaching the intended actor node, thereby increasing the risk of TTL timeouts.

AODV has the highest PDR. As can be seen from the graph, all the three protocols show a consistent PDR as we vary the simulation time while keeping the number of nodes same. DSDV has the lowest delivery data rate when a network is populated by a large number of nodes. So, in applications with fewer scalability issues, one can fairly use all the three protocols but AODV is preferred due to its high PDR.

6.21.2 Average End-to-End Delay Based Performance Evaluation

Figures 6.6 and 6.7 show the change of average end-to-end delay with the variation of number of actor nodes and simulation times, respectively, for various routing protocols.

AODV has less end-to-end delay when compared to DSDV and DSR. The performance of AODV and DSR remains consistent when the number of nodes is less, but the end-to-end delay increases slightly as the node increases. DSDV has the worst end-to-end delay as the number of node increases and performs badly when compared to AODV and DSR. In DSDV, if a node needs to have information about a destination node, then it can be obtained only

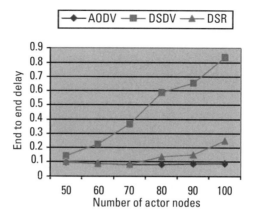

Figure 6.6 End-to-end delay vs. number of actor nodes [107].

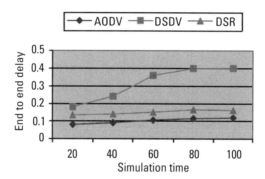

Figure 6.7 End-to-end delay vs. simulation time.

when its table gets updated by a message sent to the same destination actor node resulting in delay.

As can be seen from the graph, DSDV has the highest end-to-end delay and is the worst performer. End-to-end delay of all the protocols increase as we increase the simulation time, but it is less in DSR and AODV when compared to DSDV. AODV has less end-to-end delay. AODV provides a good balance of high PDR and low end-to-end delay, which works well for supporting higher priority traffic such as voice and video.

6.21.3 Control Packet Overhead Based Performance Evaluation

The variation of control packet overhead with the change of number of actor nodes simulation time are shown in Figures 6.8 and 6.9, respectively, for various routing protocols.

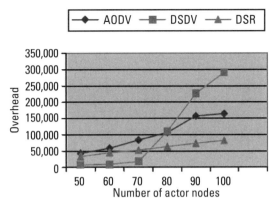

Figure 6.8 Overhead vs. number of actor nodes [107].

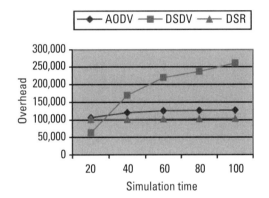

Figure 6.9 Overhead vs. simulation time.

From the graph, we can analyze that DSDV as having more overhead when compared to AODV and DSR as the number of nodes increases. The more the overhead, the less the protocol is scalable. When the number of nodes is less, DSDV has less overhead. But it tends to increase as the number of nodes increases, since the updates are propagated throughout the network in order to maintain an up-to-date view of the network topology at all the nodes. AODV produces less overhead than DSDV, as AODV tries to discover a routing path only when it is needed. DSR has less overhead than other routing protocols, as it makes use of caching mechanism and is more likely to find the routes in its cache that results in fewer route discovery requests than other protocols. Due to low overhead, DSR is more scalable than other protocols, which enables us to use DSR in applications where scalability is needed.

DSR has less overhead when the simulation time is varied while keeping the number of nodes constant. AODV has slightly higher overhead than DSR.

DSDV has the highest overhead issues due to the extensive and regular updates of the routing tables at the nodes. As the simulation time increases, the routing overhead increases, since more table updates are being sent. In AODV multiple route reply packets are sent in response to a single route request packet, leading to control overhead. This explains why routing overhead of AODV is more than DSR.

6.21.4 Throughput Based Performance Evaluation

Figures 6.10 and Figure 6.11 represent the variation of network throughput with the variation of number of actor nodes and simulation time, respectively, for some routing protocols.

Both DSR and AODV fare well. If the number of nodes is less, AODV has highest throughput. But as the number of nodes increases, both AODV and DSR have more or less the same throughput, as can be seen from the graph. DSDV has fewer throughputs, which can be attributed to the excessive channel usage by the regular route table updates. Hence, we conclude that since bandwidth is a critical issue in WSAN, we consider AODV the routing protocol for bandwidth-constrained applications.

Here both DSR and AODV have high throughput, while DSDV has less throughput. But as the number of nodes increases, both DSR and AODV start to drop in throughput. This is because as the simulation time increases, more routing traffic will be generated, as AODV and DSR use flooding for route discovery, leading to a decrease in throughput. Even though DSDV has fewer throughputs initially, it consistently increases as the simulation time increases. Because in DSDV the nodes issue routing table updates periodically, almost independent of route changes, the throughput is virtually not affected

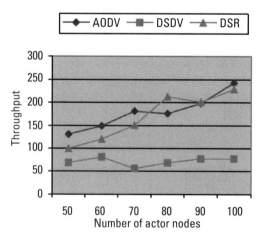

Figure 6.10 Throughput vs. number of actor nodes [107].

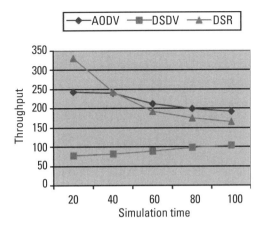

Figure 6.11 Throughput vs. simulation time.

by variation in the simulation time. In applications like flood detection, where initial throughput is more important, we can deploy AODV or DSR as an actor-to-actor protocol, but if we need a consistent throughput we can deploy DSDV provided that the number of actor nodes are not scaled frequently.

References

[1] Royer, E. M., and C. K. Toh, "A Review of Current Routing Protocols for Ad Hoc Wireless Mobile Networks," *IEEE Personal Communications*, April 1999, pp. 46–55.

[2] Akylidiz, W., S. Sankarasubramaniam, and E. Cayrici, "A Survey on Sensor Networks," *IEEE Communications Magazine*, Vol. 40, No. 8, August 2002, pp.102–114.

[3] Akkaya, K., and M. Younis, "A Survey of Routing Protocols in Wireless Sensor Networks," *Elsevier Ad Hoc Network Journal*, 2005, pp. 325–349.

[4] Jiang, Q., and D. Manivannan, "Routing Protocols for Sensor Networks." In *Proceedings of Consumer Communications and Networking Conference*, January 5–8, pp. 93–98.

[5] Parachuri, V., S. Basavaraju, A. Durresis, and R. Kannan, "Random Asynchronous Wake-up Protocol for Sensor Networks." In *Proceedings of BroadNets'04*, San Jose, CA, October 2004.

[6] Desai A., et al., "Delay-Energy Aware Routing for Sensor and Actor Networks." In *Proceedings of the 2005 11th International Conference on Parallel and Distributed Systems (IC-PADS'05)*, 2005.

[7] Peng, H., et al., "ECLS: An Efficient Cooperative Localization Scheme for Wireless Sensor and Actor Networks," In *Proceedings of the Fifth International Conference on Computer and Information Technology (CIT'05)*, 2005.

[8] Ozake, K., K. Watanabe, et al., "A Fault-Tolerant Model for Wireless Sensor-Actor System." In *Proceedings of the 20th International Conference on Advanced Information Networking and Applications (AINA'06)*, 2006.

[9] Xu, N., "A Survey of Sensor Network Applications," http://enl.usc.edu/~ningxu/paper-survey.pdf, last accessed 2010.

[10] Culler, D., D.Estrin, and M. Srivastava, "Overview of Sensor Networks," *IEEE Computer Society*, August 2004.

[11] Vieira, M., et al., "Survey on Wireless sensor Network Devices." In *Proceedings of Emerging Technologies and Factory Automation*, Vol. 1, September 16–19, 2003, pp. 537–544.

[12] http://computer.howstuffworks.com/mote1.htm.

[13] Zhang, P., M. Sadler, A. Lyon, and M. Martonosi, "Hardware Design Experiences in Zebranet," in Proceedings of SynSys '04, Baltimore, MD, November 3–5, 2004.

[14] Pottie, G., and W. J. Kaiser, "Wireless Integrated Network Sensors," *ACM Communications*, Vol. 43, No. 5, 2000, pp. 51–58.

[15] Puccinelli, D., and M. Haenggi, "Wireless Sensor Networks: Applications and Challenges of Ubiquitous Sensing," *IEEE Circuits and Systems Magazine*, Vol. 5, No. 3, 2005, pp. 19–31.

[16] Hill, J., M. Horton, R. Kling, and L. Krishnamurthy, "The Platforms Enabling Wireless Sensor Networks," *Communications of the ACM*, Vol. 47, No. 6, June 2004.

[17] Sinha, A., and A. Chandrakasan, "Dynamic Voltage Scheduling Using Adaptive Filtering of Workload Traces," In *Proceedings of the 11th International Conference on VLSI Design*, 2001.

[18] Chandrakasan, A. P., et al., "An Architecure for a Power-Aware Distributed Microsensor Nodes," IEEE Workshop on Signal Processing Systems (SiPS'00), Lafayette, LA, October 2000.

[19] Eswaran, A., A. Rowe, and R. Rajkumar, "Nano-RK: An Energy Aware Resource Centric RTOS for Sensor Networks," in Proceedings of the 26th IEEE International Real-Time Systems Symposium (RTSS'05) 2005, pp. 256–265.

[20] http://www.tinyos.net, last accessed March 2010.

[21] Levis, P., S. Madden, D. Gay, J. Polastre, and R. Szewczyk, et al., "The Emergence of Networking Abstractions and Techniques in tinyos." In *Proceedings of the First USENIX/ACM Symposium on Networked Systems Design and Implementation*, NSDI, 2004.

[22] Gay, D., P. Levis, R. von Behren, M. Welsh, and E. Brewer et al., "The nesc Language: A Holistic Approach to Networked Embedded Systems." In *Proceedings of the ACM SIGPLAN 2003 Conference on Programming Language Design and Implementation*, San Diego, CA, 2003.

[23] Abrach, H., S. Bhatti, et.al., "MANTIS: System Support for Multimodal Networks of In-Situ Sensors," 2nd ACM International Workshop on Wireless Sensor Networks and Applications, 2003, pp. 50–59.

[24] Silberchatz, A., P. B. Galvin, and G. Gagne, *Operating Systems*, John Wiley Publication, 7th edition, 2006.

[25] Park, S., J. W. Kim, et al., "Embedded Sensor Networked Operating System." In *Proceedings of the Ninth IEEE International Symposium on Object and Component-Oriented Real-Time Distributed Computing*, 2006, pp.117–124.

[26] Wong, K. D., "Physical Layer Considerations for Wireless Sensor Networks," IEEE International Conference in Network Sensing and Control, March 2004, pp. 1201–1206.

[27] Lin, T.-H., W. J. Kaiser, and G. J. Pottie, "Integrated Low Power Communication System Design for Wireless Sensor Networks," *IEEE Communications*, 2004, pp. 142–150.

[28] Zhou, Y., J. Xing, and Q. Yu, "Overview of Power-Efficient MAC and Routing Protocols for Wireless Sensor Networks." In *Proceedings of the 2nd IEEE/ASME International Conference on Mechatronics and Embedded Systems and Applications*, August 2006, pp. 1–6.

[29] Czapski, P. P., "A Survey: MAC Protocols for Applications of Wireless Sensor Networks." In *Proceedings of TENCON 2006*, Hong Kong, November 2006, pp. 1–4.

[30] Chiras, T., M. Paterakis, and P. Koutsakis, "Improved Medium Access Control for Wireless Sensor Networks: A Study on the S-Mac Protocol." In *Proceedings of the 14th IEEE Workshop on Local and Metropolitan Area Networks*, LANMAN, 2005.

[31] Ye, W., J. Heidemann, and D. Estrin, "An Energy-Efficient MAC Protocol for Wireless Sensor Networks," *IEEE Infocomm*, 2002.

[32] Demirkol, I., C. Ersoy, and F. Alagoz, "MAC Protocols for Wireless Sensor Networks: A Survey," *IEEE Communications*, April 2006.

[33] Warrier, A., et al., "Mitigating Starvation in Wireless Sensor Networks," Military Communications Conference, 2006, pp. 1–5.

[34] Lin, R., Z Wang, and Y. Sun, "Energy Efficient Medium Access Controls for Wireless Sensor Networks and Its State of Art," IEEE, 2004.

[35] Dantu, K., M. Rahim, et al., "RoboMote: Enabling Mobility in Sensor Networks," IEEE/ACM Fourth International Conference on Information Processing in Sensor Networks (IPSN/SPOTS), April 2005.

[36] Ringwald, M., and K. Romer, "Deployment of Sensor Networks: Problems and Passive Inspection." In *Proceedings of Fifth International Workshop on Intelligent Solutions in Embedded Systems*, Madrid, Spain, 2007.

[37] Ahmed, A., et al., "Wired vs. Wireless Deployment Support for Wireless Sensor Networks," IEEE Region 10 Conference, TENCON, 2006, pp. 1–3.

[38] Li, J., Y. Bai, H. Ji, and D. Qian, "POWER: Planning and Deployment Platform for Wireless Sensor Networks." In *Proceedings of the Fifth International Conference on Grid and Cooperative Computing Workshops (GCCW'06)*, IEEE, 2006.

[39] Zhenjie, X., and C. Changjia, "A Localization Scheme with Mobile Beacon for Wireless Sensor Networks." In *Proceedings of International Conference on ITS Telecommunications Proceedings*, 2006.

[40] Patro, R. K., "Localization in Wireless Sensor Networks with Mobile Beacons," 23rd IEEE Convention of Electrical and Electronics Engineers in Israel, 2004, pp. 22–24.

[41] Savvides, A., C. C. Han, and M. B. Srivastava, "Dynamic Fine Grained Localization in Ad Hoc Networks of Sensors." In *Proc. of Mobicom 2001*, July 2001, pp. 166–179.

[42] Pandy, S., P. Prasad, P.Sinha, and P. Agarwal, "Localization of Sensor Networks Considering Energy Accuracy Tradeoffs." In *Proceedings of International Conference on Collaborative Computing: Networking, Applications and Worksharing*, December 19–21, 2005.

[43] Hu, L., and D. Evans,"Localization for Mobile Sensor Networks." In *Proceedings of the Tenth Annual International Conference on Mobile Computing and Networking*, Philadelphia, PA, September 26–October 1, 2004.

[44] Elson, J., and K. Romer, "Wireless Sensor Networks: A New Regime for Time Synchronization." In *Proceedings of the First Workshop on Hop Topics in Networks*, Princeton, NJ, October 28–29, 2002.

[45] Chaudhuri, S. P., A. K. Saha, and D. B. Johnson, "Adaptive Clock Synchronization in Sensor Networks," IPSN'04, Berkeley, CA, April 26–27, 2004.

[46] Sivrikaya, F., and B. Yener, "Time Synchronization in Sensor Networks: A Survey," *IEEE Network*, July/August 2004.

[47] Bychkovskiy, V., S. Megerian, D. Estrin, and M. Potknojak, "A Collaborative Approach to In Place Sensor Calibration," 2nd International Workshop on Information Processing in Sensor Networks (IPSN'03), Palo Alto, CA, April 2003, pp. 301–316.

[48] Feng, J., S. Megerian, and M. Potkonjak, "Model Based Calibration for Sensor Networks." In P*roceedings of IEEE Sensors*, Vol. 2, October 22–24, 2003, pp. 737–742.

[49] Whitehouse, K., and D. Culler, "Calibration as Parameter Estimation in Sensor Networks." In *Proceedings of WSNA'02*, Atlanta, GA, September 28, 2002, pp. 59–67.

[50] Wang, L., "Survey on Sensor Networks," Department of Computer Science Engineering, Technical Report MSU-CSE-04-19, Michigan State University 2004.

[51] Ganesan, D., et al., "Networking No.s in Wireless Sensor Networks," *Elsevier Science*, December 9, 2005.

[52] Jiang, P., and Y. Wen, et al., "A Study of Routing Protocols in Wireless Sensor Networks." In *Proceedings of the 6th World Congress on Intelligent Control and Automation*, Dalian, China, June 21–23, 2006.

[53] Heinzelman, W. R., A. Chandrakasan, and H. Balakrishnan, "Energy-Efficient Communication Protocol for Wireless Microsensor Networks." In *Proc. IEEE Hawaii Int'l Conference*, January 2000, pp. 1–10.

[54] Govindan, R., and D. Estrin, "Directed Diffusion: A Scalable and Robust Communication Paradigm for Sensor Networks." In *Proceedings of the Sixth Annual ACM/IEEE International Conference on Mobile Computing and Neworking(MOBICOM 2000)*, August 2000, Boston MA: ACM Press, pp. 56–67.

[55] Van Dinh, D., et al., "Wireless Sensor Actor Networks and Routing Performance Analysis." In *Proceedings of International Workshop on Wireless Ad Hoc Networks*, London, UK, 2005.

[56] Wang, C., et al., "A Survey of Transport Protocols for Wireless Sensor Networks," *IEEE Network*, Vol. 20, No. 3, June 2006, pp. 34–40.

[57] Mansouri, V. S., and B. Afsari, "A Simple Transport Protocol for Wireless Sensor Networks." In *Proceedings of ISSNIP*, 2005.

[58] Wan, C.-Y., and L. Krishnamurthy, "Pump-Slowly, Fetch-Quickly (PSFQ): A Reliable Transport Protocol for Sensor Networks," *IEEE Journal on Selected Areas in Communications*, Vol. 23, No. 4, April 2005.

[59] Rajagopalan, R., and K. P. Varshney, "Data Aggregation Techniques in Sensor Networks: A Survey," *IEEE Communications Surveys and Tutorials*, Fourth Quarter, 2006.

[60] Li, X., "A Survey on Data Aggregation in Wireless Sensor Networks," Project Report for CMPT 765, Spring 2006.

[61] Duarte-Melo, E. J., and M. Liu, "Data Gathering Wireless Sensor Networks: Organization and Capacity," *Computer Networks*, Vol. 43, 2003, pp. 519–537.

[62] Boulis, A., S. Ganeriwal, M. B. Srivastava, "Aggregation in Sensor Networks: An Energy Accuracy Trade-Off," *Elseiver's Ad Hoc Networks*, Vol. 1, 2003, pp. 317–331.

[63] Tilak, S., A. Ghzaleh, and W. Heinzelman, "A Taxonomy of Wireless Micro Sensor Network Models," *Mobile Computing and Communication Review*, Vol. 6, No. 2, 2002.

[64] Zhang, W., G. Cao, and T. La Porta, "Data Dissemination with Ring Based Index for Wireless Sensor Networks," *IEEE Transactions on Mobile Computing*, Vol. 6, No. 7, July 2007.

[65] Zhuang, L. Q., J. B. Zhang, D.H. Zhang, and Y. Z. Zhao, "Data Management for Wireless Sensor Networks: Research No.s and Challenges." In *Proceedings of 2005 International Conference on Control and Automation (ICCA2005)*, Budapest, Hungary, June 27–29, 2005.

[66] Bonnet, P., J. Gehrke, and P. Seshadri, "Querying the Physical World," *IEEE Personal Communications*, October 2000.

[67] Navas, J. C., and M. Wnyblatt, "The Network Is the Database: Data Management for Highly Distributed Systems," ACM SIGMOD, Santa Barbara, CA, May 21–24, 2001.

[68] Deshpande, A., C. Guestrin, et al., "Model-Driven Data Acquisition in Sensor Networks." In *Proceedings of the 30th VLDB Conference*, Toronto, Canada, 2004.

[69] Govendan, R., J. M. Hellerstein, W. Hong, et al., "The Sensor Network as a Database," Technical Report 02-771, Computer Science Department, University of Southern California, September 2002.

[70] Ratnasamy, R., R. Govindan, et al., "Data Centric Storage in Sensornets with GHT, A Geographic Hash Table," *Mobile Network and Applications (MONET)*, Special No. on Wireless Sensor Networks, Kluwer, 2003.

[71] Bose, P., P. Morin, J. Urrutia, "Routing with Guaranteed Delivery in Ad Hoc Wireless Networks," Wireless Networks, 7 (6) (2001) 609–616, DealM, 00, Seattle, Washington, August 1999, pp. 45–48.

[72] Yo, Y., and J. Gehrke, "Query Processing for Sensor Networks." In *Proceedings of the 2003 CIDR Conference*, 2003.

[73] Bonnet, P., J. Gehrke, and P. Seshadri, "Querying the Physical World,"IEEE Personal Communications, October 2000.

[74] Madden, S. R., M. J. Franklin, J. M. Hellerstein, and W. Hong, "TinyDB: An Acquisitional Query Processing System for Sensor Networks," *ACM Transactions on Database Systems*, 2004.

[75] http://telegraph.cs.berkeley.edu/tinydb, last accessed February–March 2010.

[76] http://cougar.cs.cornell.edu, last accessed 2010.

[77] Sugihara, R., and R. K. Gupta, "Programming Models for Sensor Networks: A Survey," *ACM Transactions on Sensor Networks*, 2006.

[78] Choeng, E., J. Liebman, J. Liu, and F. Zhao, "TinyGALS: A Programming Model for Event-Driven Embedded Systems." In *Proc. SAC'03*, pp. 698–704.

[79] Bischoff, U., and G. Kortuen, "A State Based Programming Model and System for Wireless Sensor Networks." In *Proceedings of the Fifth Annual IEEE International Conference on Pervasive Computing and Communicatins Workshops (PerComW'07)*, 2007.

[80] Cheong, E., and J. Lie, "galsC: A Language for Event-Driven Embedded Systems." In *Proceedings of the Design, Automation and Test in Europe Conference and Exhibition (DATE'05)*, IEEE, 2005.

[81] Hadim, S., N. Mohamed, "Middleware: Middleware Challenges and Approaches for Wireless Sensor Networks," *IEEE Distributed Systems Online 1541–4922*, IEEE Computer Society, Vol. 7, No. 3, March 2006.

[82] Henricksen, K., and R. Robinson, "A Survey of Middleware for Sensor Networks: State of the Art and Future Directions." In *Proceedings of MidSens'06*, Melbourne, Australia, November 27–December 1, 2006.

[83] Wolenetz, M., "Middleware Guidelines for Future Sensor Networks," First Annual International Conference on Broadband Networks, San Jose, CA, October 25–29, 2004.

[84] Romer, K., O. Kasten, and F. Mattern, "Middleware Challenges for Wireless Sensor Networks," *Mobile Computing and Communications Review*, Vol. 6, No. 2. {AU: Year?}

[85] Yu, Y., B. Krishnamachari, and V. K. Prasanna, "No.s in Designing Middleware for Wireless Sensor Networks," *IEEE Network*, January–February, 2004.

[86] Hadim, S., and N. Mohammed, "Middleware for Wireless Sensor Networks," International Conference on Parallel Processing Workshops, August 14–18, 2006.

[87] Molla, M. M., and S. I. Ahmad, "A Survey of Middleware for Sensor Network and Challenges." In *Proceedings of the 2006 International Conference on Parallel Processing Workshops (ICPPW'06)*, IEEE 2006.

[88] Levis, P., and D. Culler, "Mate: A Tiny Virtual Machine for Sensor Networks." In *Proceedings of the 10th International Conference on Architectural Support for Programming Languages and Operating Systems*, San Jose, CA, October 2002.

[89] Boulis, C., C. Han, and M. B.Srivastava, "Design and Implementation of a Framework for Programmable and Efficient Sensor Networks," MobiSys 2003, San Francisco, CA, May 2003.

[90] Chen, D., and P. K. Varshney, "QoS Support in Wireless Sensor Networks: A Survey." In *Proceedings of the 2004 International Conference on Wireless Network (ICWN 2004)*, Las Vegas, NV, June 21–24, 2004.

[91] Younis, and M., K. Akkaya, et al, "On Handling QoS traffic in Wireless Sensor Network." In *Proceedings of the 37th Hawaii International Conference on System Science*, 2004.

[92] Kay, J., and J. Frolik, "Quality of Service Analysis and Control for Wireless Sensor Networks," IEEE International Conference on Mobile Ad-Hoc and Sensor Systems, 2004.

[93] Wang, Y., X. Liu, and J. Yin, "Requirements of Quality of Service in Wireless Sensor Networks." In *Proceedings of the International Conference on Mobile Communications and Learning Technologies (ICNICONSMCL'06)*, IEEE, 2006.

[94] Xing, L., and A. Shrestha, "QoS Reliability of Hierarchical Clustered Wireless Sensor Networks." In *Proceedings of 25th IEEE International Conference on Performance, Computing and Communications*, April 10–12, 2006.

[95] Shhrabi, K., J. Gao, V. Ailawadhi, and G. J. Pottie, "Protocols for Self Organization of a Wireless Sensor Networks," *IEEE Personal Communications*, October 2000, pp. 16–27.

[96] Pathan, A.-S. K., H.-W. Lee, and C. S. Hong, "Security in Wireless Sensor Networks: No.s and Challenges," IEEE ICACT, 2006.

[97] Zia, T., and A. Zomaya, "Security No.s in Wireless Sensor Networks." In *Proc. International Conference on Systems and Networks Communication (ICSNC'06)*, 2006, p. 40.

[98] Wang, Y., G. Attebury, and B. Rammurthy, "A Survey of Security No.s in Wireless Sensor Networks," *IEEE Communications Survey*, Vol. 8, No 2, 2006.

[99] Djenouri, D., L. Khelladi, and N. Badache, "A Survey of Security No.s in Mobile Ad Hoc and Sensor Networks," *IEEE Communications Surveys*, Vol. 7, No. 4, 2005.

[100] Shi, E., and A. Perrig, "Designing Secure Sensor Networks," *IEEE Wireless Communications*, Vol. 11, No. 6, December 2004, pp. 38–43.

[101] Guimaraes, G., E. Souto, and D. Sadok, "Evaluation of Security Mechanisms in Wireless Sensor Networks." In *Proceedings of the Systems Communications (ICW'05)*, 2005.

[102] Xiangyu, J., and W. Chao, "The Security Routing Research for WSN in the Application of Intelligent Transport System." In *Proceedings of the 206 IEEE International Conference on Mechatronics and Automation*, Luoyang, China, June 25–28, 2006.

[103] Conti, M., S. Giordano, G. Maselli, and G. Turi, "Cross-Layering in Mobile Ad-Hoc Network Design," *IEEE Computer*, Vol. 37, No. 2, 2004, pp. 48–51.

[104] Akylidiz, I. F., and I. H. Kasimoglu, "Wireless Sensor and Actor Networks: Research Challenges," *Ad Hoc Networks Journal*, Vol. 2, No. 4, October 2004, pp. 351–367.

[105] Kurkowski, S., T. Camp, and M. Colagrosso, "MANET Simulation Studies: The Incredibles," Vol. 9, No. 4, October 2005.

[106] NRL's Sensor Network Extension to NS-2, http://www.nrlsensorsim.pf.itd.nrl.navy.mil/, last accessed 2010.

[107] Proceedings of International Conference on Computer Networks and Security (ICCNS08), September 27–28, 2008, Pune, India.

7

Performance Analysis of IEEE 802.15.4 with Sink Nodes in WSN Scenario

7.1 Introduction

Wireless sensor networks have become a hot research theme in academia and as well as in industry in recent years due to their wide range of applications. By appropriately tuning the parameters of IEEE 802.15.4, it can be applied to a variety of applications. Research on sensor networks has been stimulated by the need for setting up communication networks to gather information in situations where fixed infrastructure cannot be employed on the fly, as it occurs in the management of emergencies and disaster recovery [1–3].

In a sensor network, thousands of sensor nodes are deployed in a random fashion. The sensor nodes sense the phenomenon periodically, and the sensed data is sent to the sink node. The information collected at the sink node is queried to extract the relevant information. In sensor networks, by shortening the distance taken by the packets to reach the sink node, energy can be conserved. Mobility of sink may result in retrieving the data quickly [4].

In this chapter, we study the effect of the IEEE 802.15.4 MAC standard on the performance of AODV in a sensor environment with static and mobile ad hoc sink nodes. The IEEE 802.15.4 is the standard adopted for wireless sensor network platform. AODV is defined to be used as the underlying routing protocol in ZigBee.

The sensor network should be pliant enough in nature to allow a systematic deployment of sensor nodes, including mobility among the sink nodes. The disseminated data from the sensed nodes gather at the sink node. Data dissemination is the major source for energy consumption in a sensor network. Sensor nodes near the sink disperse a large amount of data to the sink with less energy consumption, while the nodes far away from the sink require more energy to do the same. Hence, there is a need to investigate to see whether the dispersed data can be assimilated in vast quantities by moving the sink node to the region where large number of sensor nodes is emitting the sensed data. This leads to the question, what is the maximum speed at which the sink nodes need to be moved? The mentioned scenario is simulated using NS2 with WPAN extension, which is an open source network simulator tool, and from our simulative evaluation we show that the sink node velocity should be less than 1m/s for obtaining acceptable performance.

The main contributions of this chapter are as follows:

- A comprehensive analysis of the effect of the IEEE 802.15.4 MAC protocol on the performance of the ZigBee routing protocol is evaluated with static nodes in wireless sensor networks.
- Substantial amount of effort has been put to analyze the effect of IEEE 802.15.4 MAC protocol on the performance of AODV routing protocol in wireless sensor networks with mobile ad hoc sink nodes. There has been no substantial work reported in evaluating IEEE 802.15.4 from an ad hoc mobile sink point of view.

7.2 Literature Background

The IEEE 802.15.4 standard was implemented by J. Zheng et al. [5] on the NS2 simulator, and they carried out a comprehensive study of the 802.15.4 standard. The authors have conducted the simulation in beacon and nonbeacon-enabled mode. The authors also have tested various features like association, tree formation, network auto-configuration, orphaning, and coordinator relocation.

Performance comparison of IEEE 802.15.4 and IEEE 802.11 is conducted in Qicai Yu et al. [6] for low-rate personal area networks. They have considered various parameters like bandwidth efficiency, packet delivery, average delay, and energy consumption. It has been deduced that the bandwidth efficiency of IEEE 802.15.4 increases as the MSDU size increases. A steep fall in packet delivery of IEEE 802.15.4 can be observed as the pkts/sec increases. Also, energy consumption in IEEE 802.11 is more when compared to IEEE 802.15.4.

IEEE 802.15.4 is relatively a new technology, and products based on IEEE 802.15.4 are yet to be realized. But products based on IEEE 802.11 have made significant impact on our day-to-day applications. Lewis Adams [7] discusses sensor network scenarios using 802.11. It's proposed that by tuning the parameters of IEEE 802.11, it can be made energy efficient. Energy efficient 802.11 can be readily applied in a sensor environment.

In Pore et al. [8], OLSR is compared with other protocols like AODV and DSDV in an oceanic sensor network environment. Parameters like packet delivery ratio, average end-to-end delay, routing overhead, and normalized routing load are considered for comparison. These routing protocols were compared both in static mode and in mobility mode. The authors have shown that OLSR performs better than the AODV and DSDV in every scenario.

Another implementation of IEEE 802.15.4 on NS2 was carried out by Lu et al. [9]. The authors have evaluated the star topology network scenario with a beacon-enabled mode. The authors conclude that an extremely low-duty cycle operation enables significant energy savings, but these savings come at the cost of high latency and low bandwidth.

In Gomez et al. [10], the authors have evaluated the possibility of adopting AODV routing protocol over IEEE 802.15.4 in a mesh sensor network. The authors have proposed a new version of AODV routing protocol called NST-AODV specially designed for sensor networks. Different implementations of AODV like AODVjr, AODVbis, LoWPAN-AODV LOAD, and Tiny AODV have been discussed and compared with NST-AODV. It has been shown that NSTAODV reduces network delay and the number of retransmissions of the packets while increasing the network reliability.

The IEEE 802.15.4 based sensor network is evaluated with different topologies in Hoffert et al. [11]. The topologies considered are ideal, fully connected topology; nonideal fully connected topology; ideal star network topology; and nonideal star network topology. Experimental setup is conducted in beacon and nonbeacon-enabled mode. The following recommendations are provided by considering the amount of data delivered in each experimental setup. Firstly, if a node is not directly associated with a PAN coordinator, then for such nodes the GTS should be enabled. Secondly, flexibility is required in configuration of slots in a superframe, and lastly, even if a node is not associated with a PAN coordinator, messages can still be sent to those nodes by providing beacons to the PAN coordinator.

Under the CSMA-CA mechanism and beacon-enabled mode, the IEEE 802.15.4 has been investigated by considering data payload size and direct and indirect data transmissions [12]. Various concepts like function devices, network topology, superframe structures, and CSMA-CA mechanisms have been discussed. Practical study of IEEE 802.15.4 has been conducted using CC2420 Chipcon devices with delivery ratio, throughput, and RSSI as the metrics.

By default the number of backoffs declared in CSMA-CA mechanism of IEEE 802.15.4 is four. But in reality, the values supported ranges from one to five. The authors have modified the default value and have evaluated the effect of having low backoff on 802.15.4. By simulation the authors have shown that less backoff mechanism leads to less power consumption and less latency (Antonis et al. [13]). In Ko et al. [14], IEEE 802.15.4 has been evaluated with different backoff values, and they have also proposed a "state transition scheme." Here, the minBE values are changed dynamically based on transmission schemes to make successful transmissions. In the state transmission scheme, a starting value of 3 (minBE) is considered for the nodes. But once there is data to be transferred, then the value is changed to a lower value like 2. If the data transmission is successful, then again minBE is reduced to 1. If there is no data to be transmitted, then again minBE is changed to 2. By applying this state transmission technique, high throughput was achieved.

Evaluation of IEEE 802.15.4 with mobile sink is done in Chen et al. [15]. But the main difference between our work and the work in [15] is that they have not obtained the results by varying the mobility speed of the sink node against various traffic load, number of sources, and number of nodes.

7.3 Overview of IEEE 802.15.4 and Its Characteristics

IEEE 802.15.4 is a low-rate personal wireless area network standard. IEEE 802.15.4 works in three different frequency bands: 868-MHz band working at a data rate of 20 Kbps, 915-MHz band working at a data rate of 40 Kbps, and 2.4-GHz band working at a data rate of 250 Kbps. Different topologies supported in IEEE 802.15.4 are star topology or peer-to-peer network or a cluster tree network. In IEEE 802.15.4 standard, 14 PHY and 35 MAC primitives have been defined. A device can be either a fully function device (FFD) or a reduced function device (RFD). Any device that is not a co-coordinator is an end node. FFD can act as a PAN coordinator, a coordinator, or just as an end node (device). FFD also functions as a routing device for grid topologies and for peer-to-peer communications. RFD has a reduced set of functionalities, which can only function as an end device or node. It does not have the ability to communicate with any device other than the coordinator. A simplistic sensor network is made up of a mixture of these devices. But the basic rule is that any PAN network should have at least one FFD to act as the PAN coordinator or a sink node. Major portions of the text in this section including figures are quoted from the IEEE 802.15.4 Standard Specification Manual [16].

7.3.1 Network Topologies

Two types of topologies are supported in LR-WPAN. They are star topology (Figure 7.1(a)) and peer-to-peer topology (Figure 7.1(b)). A star topology consists of a central controller called the PAN coordinator [12]. This PAN coordinator establishes communication with the remaining devices. The other devices communicate with the PAN coordinator using a 64-bit extended address. In star topology, communication occurs in a one-to-one fashion between the devices and the PAN coordinator. The other devices do not communicate among themselves. The PAN coordinator is found in the peer-to-peer technology also. Unlike the star topology the devices in a peer-to-peer topology can communicate among themselves and also with the PAN coordinator.

Network formation is carried out using these topologies. A network can consist of numerous star topologies with each of the star topologies operating independently without any conflict. This is possible through a device called a *PAN identifier*. The PAN identifier makes sure that the FFD and RFD join the star topology in an appropriate manner. One of the best examples of the peer-to-peer topology is the cluster tree network. Any FFD that communicates with the channel is selected as the PAN coordinator. The cluster tree network is mainly made up of FFDs, but the network may consist of RFDs also. Synchronization is done through the use of coordinators. The PAN coordinator declares itself the cluster head with a cluster identification value of zero. It chooses an unusual FFD as a PAN identifier. In order to establish a cluster, the PAN coordinator that had declared itself the cluster head broadcasts the beacon to the neighboring devices. A FFD receiving the beacon will then request to join the cluster head. If the PAN coordinator gives permission, then the device may join the cluster head as its child. The cluster head is added as the parent device in the list of the child node. In this way many clusters can be formed, with each cluster

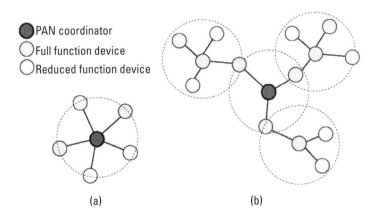

Figure 7.1 (a) Star and (b) cluster tree topology of IEEE 802.15.4 network [12].

having a cluster head. The cluster head for a new cluster is selected by the PAN coordinator itself. This type of network results in greater coverage of area while increasing the delay and overhead in the network.

7.3.2 LR-WPAN Protocol Architecture

The IEEE 802.15.4 device architecture consists of MAC layer, PHY layer, and upper layer (Figure 7.2). The MAC layer provides connectivity to the physical medium. They PHY layer consists of radio frequency transceiver [17]. The main function of PHY layer is to switch ON and OFF of the radio devices, sensing the channel, link quality detection, and selecting of channels. Energy is conserved by toggling the radio devices in ON and OFF mode. Sensing of channels is carried out, so that a free channel can be allocated for communication. Energy detection (ED) and clear channel assessment (CCA) are carried out to see whether a channel is free or not. Assessment of link quality is necessary, as it affects the performance of communication between various devices. The MAC layer is responsible for the generation of beacons, synchronization of various devices using the beacons, association, and disassociation of devices with the coordinator, using CSMA-CA for accessing the channel, maintenance of GTS mechanisms, and to act as a bridge between any two peer MAC entities. The upper layer provides routing, network configuration, and management. The

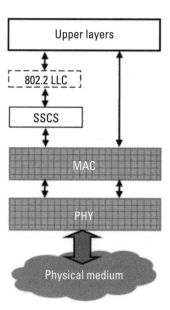

Figure 7.2 LR-WPAN device architecture [17].

service specific convergence sublayer (SSCS) acts as an intermediate between logical link control (LLC) and the MAC layer for communication occurrence.

7.3.3 The Superframe Structure

An IEEE 802.15.4 can be operated both in a beacon and nonbeacon-enabled mode. In a beacon-enable mode, the pan coordinator sends a special frame called a beacon for synchronization with other nodes. In a nonbeacon-enabled mode, there are no regular beacons and communication is carried out using unslotted CSMA [18].

If a device wants to communicate during the contention access period (CAP), then it has to compete with other devices using CSMA-CA mechanisms. The PAN coordinator may allocate a part of the active portion of the superframe to some applications. This portion is called a guaranteed time slot (GTS). This GTS is part of the contention free period (CFP). Seven such GTS slots can be allocated by the PAN coordinator. Still the remaining part of the active super frame shall be accessible for other devices.

A superframe structure in a beacon-enabled mode is as shown in Figure 7.3. The super frame has a beacon on either side of the structure. An active and an optional inactive period follow the starting beacon. During the active period, communication takes place. The superframe structure is described by the value of macBeaconOrder and macSuperframeOrder. The time interval at which the coordinator transmits the beacon frames is given by the macBeaconOrder (BO). The macBeaconOrder (BO) and BeaconInterval (BI) are related by BI = (aBaseSuperframeDuration2^{BO}) such that 0 BO 14. The length of the active portion of the superframe is given by macSuperframeOrder. The

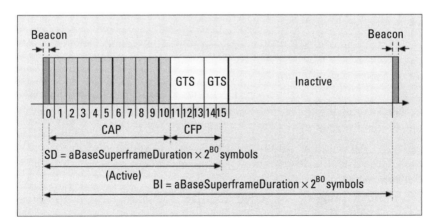

Figure 7.3 Superframe structure of IEEE 802.15.4 in a beacon-enabled mode [18].

relation between macSuperframeOrder (SO) and superframe duration (SD) can be described as SD = (aBaseSuperframeDuratio2^{SO}) where 0 SO 14. If the value of SO is 15, then it will become inactive. A value of 15 for BO indicates that the superframe is nonexistent. When macBeaconOrder and macSuperframeOrder is set to 15 then the PAN operates in a nonbeacon-enabled mode. If a PAN has to operate in a beacon-enabled mode, then macBeaconorder and macSuperframeOrder values should be set between 0 and 14. The active portion of the superframe is divided into equally spaced slots. The duration of each slot is 2SO * aBaseSlotDuration. Each active period consists of a beacon, CAP, and CFP. A beacon occupies the 0 slot and is transmitted irrespective of CSMA. Transmission of beacon is followed by CAP, which is again followed by CAP if there is any, and it extends until the end of the active part. The IFS period is used by the MAC sub layer to process the data received by the PHY [20–22].

7.3.4 Data Transmission

Data transmission can occur from coordinator to device and from device to coordinator. Coordinator-to-device data transmission in a beacon-enabled mode is as follows (Figure 7.4(a)). If there is any pending data, then it is indicated by the coordinator in the beacon. Then the device listens for the beacon. If there is any data pending, then the device requests for the data through the slotted CSMA-CA. On receiving the request, the coordinator acknowledges the data request and sends the requested data through slotted CSMA-CA [16].

Device-to-coordinator data transmission in a beacon-enabled mode takes place as follows (Figure 7.4(b)). The device first listens for any available beacons. Then the device synchronizes itself with the superframe structure. Using slotted CSMA-CA mechanism, the data is sent to the coordinator by the device. The coordinator then sends an optional acknowledgment packet. In a nonbeacon-enabled mode, if the coordinator wants to send the data to a device, then it stores the data before transmitting. The coordinator stores the data so that the appropriate device can make a contact and request for the data. A device may contact the coordinator by sending the request to the coordinator using an unslotted CSMA-CA. The coordinator then sends an acknowledgment for successfully receiving the request from the device. If there is any data pending, then the data is sent to the device using unslotted CSMA-CA; otherwise, a zero length payload is sent indicating that no data is present. On successfully receiving the data, the device sends an acknowledgment frame to the coordinator (Figure 7.4(c)). In nonbeacon mode, the data is sent from a device to the coordinator by using the unslotted CSMA-CA (Figure 7.4(d)). An optional acknowledgment frame is sent by the coordinator to the network device on successfully receiving the data.

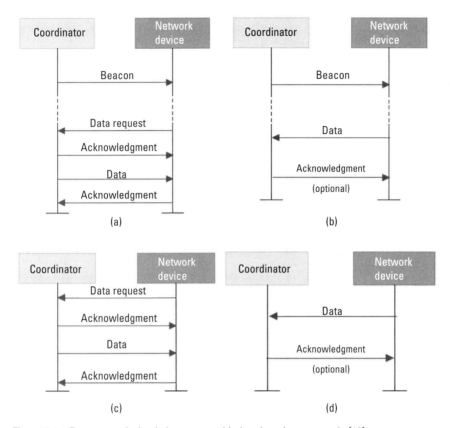

Figure 7.4 Data transmission in beacon-enabled and nonbeacon mode [16].

7.3.5 Slotted CSMA-CA Mechanism (Beacon Enabled)

The CSMA-CA algorithm uses a slotted version for a beacon-enabled mode (Figure 7.5(a)) and unslotted version for a nonbeacon-enabled mode (Figure 7.5(b)). The CSMA algorithm makes use of a concept called *backoff periods*. A backoff period is defined as a unit of time.

One backoff period is equivalent to aUnit Backoff Period symbol. In slotted CSMA-CA, the first backoff period of every device is related to the beginning of the beacon transmission. In unslotted CSMA-CA, there is no link in time between the backoff period of any of the devices. Three types of variables are supported in every transmission. They are the Number of Backoffs (NB) contention window (CW) length value, and backoff exponent (BE). The NB value determines the number of backoffs that needs to be attempted while transmitting the data frame. A value of 0 indicates a new transmission attempt. This value shows the number of backoff periods that needs to be cleared before a channel can be accessed for transmission. A default value of 2 is used before transmission and is again reset to 2 if the channel is busy. The CW variable is

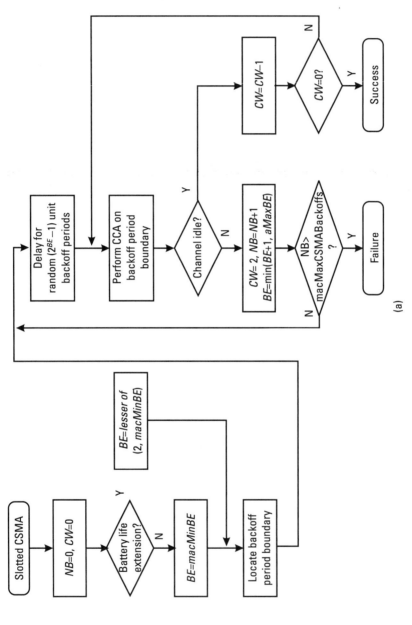

Figure 7.5 (a) Slotted CSMA-CA in (a) beacon mode, and (b) unslotted CSMA-CA in a nonbeacon mode [19].

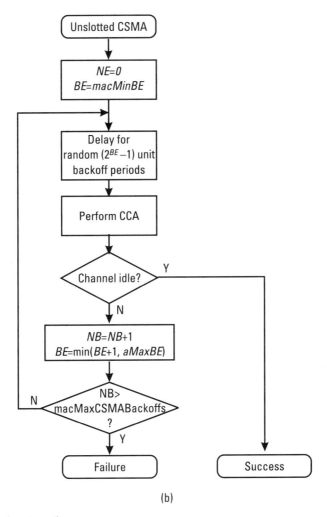

Figure 7.5 (continued)

found only in the slotted version of CSMA-CA. If the device is busy or inaccessible, then the device has to wait for a number of backoff periods. This backoff period value is decided by BE. In slotted CSMA, as a first step the CSMA-CA algorithm initializes the NB, CW, and BE variables. The end boundary of the current backoff period and beginning boundary of the next backoff period is also determined. In an unslotted CSMA-CA, only NB and BE values are initialized. In step 2, before requesting for the PHY, the MAC sublayer delays the backoff periods for a random number of backoff period in the range of 0 to $2^{BE}-1$. In step 3, the MAC sublayer requests the PHY to carry out the CCA operation. At a backoff period boundary, the commencement of CCA happens. CCA starts immediately in an unslotted CSMA-CA. A CCA operation determines whether a channel is busy or idle. If the channel is busy, then the MAC sublayer increments the value of NB and BE to one. The value of BE should not be more than aMaxBE. The value of CW is set to 2 by the slotted version of the CSMA-CA algorithm. If the value of NB is less than or equal to macMaxCSMABackoff, then the CSMA-CA algorithm proceeds to step 2. If the value of NB is greater than macMaxCSMABackoff, then the CSMA-CA algorithm terminates by displaying a channel access failure status. Accessing an idle channel is done in step 5. Before transmitting any data, the contention window value needs to be expired. This is achieved by decrementing the value of CW by one by the MAC sublayer. If the value of CW is equal to 0, then the data is transmitted at the beginning boundary of the next backoff period. If the value of CSMA-CA is not equal to 0, then the CSMA-CA algorithm proceeds to step 3. If the channel is idle in an unslotted CSMA-CA, then the data frames are transmitted immediately [23, 24].

7.3.6 Starting and Maintaining PANs

The steps involved in starting and maintaining PANs include scanning the channels to identify whether there is any conflict in the selection of PAN identifiers and finally starting up the PANs.

Different types of channel scan performed are ED channel scan, active channel scan, passive channel scan, and orphan channel scan.

- *ED channel scan:* The peak energy in each of the requested channels is measured by the FFD by performing an ED channel scan. An ED scan over a logical channel is performed by using MLME-CAN.request primitive by setting the ScanType parameter to perform an ED scan. The MLME primitive first switches to the requested channel by setting the phyCurrentChannel and performs the ED measurement for [aBaseSupterframeDuration * $(2^n + 1)$] symbols, where n is the value of the ScanDuration parameter in the MLME-Scan.request primitive. The ED scan terminates after measuring the energy in each of the specified logical channels.

- *Active channel scan:* A coordinator transmitting the beacon frames is located by a FFD by performing an active channel scan if the coordinator is within the POS of the FFD. An active scan allows a prospective PAN coordinator to select a PAN identifier to start a new PAN, or it could also be used by a FFD or RFD to get associated with the coordinator. The MLME primitive first switches to the requested channel by setting the phyCurrentChannel and sends a beacon request. The device is enabled for [aBaseSupterframeDuration * $(2^n + 1)$] symbols, where the value of n ranges from 0 to14. When the number of beacons transmitted reaches an implementation-specific value, then the active scan on the particular channel is terminated.

- *Passive channel scan:* The passive channel scan, just like active scan, allows FFD or RFD devices to get associated with a coordinator provided that the coordinator is within the range of the device but without beacon request. The MLME primitive first switches to the requested channel by setting the phyCurrentChannel for [aBaseSupterframeDuration * (2^n + 1)] symbols, where the value of n ranges from 0 to14. Just like an active scan, the passive scan is terminated when the number of beacons transmitted reaches an implementation specific value.

- *Orphan channel scan:* When a device loses synchronization with the coordinator. then it performs an orphan channel scan so that it can reconnect with a coordinator. The MLME primitive first switches to the requested channel by setting the phyCurrentChannel and sends an orphan notification command. A device receiving this notification will have aResponseWaitTime to respond. An orphan scan is terminated when a device realigns with a coordinator.

- *Resolution for PAN identifier conflict:* A PAN identifier conflict can be detected either by the PAN coordinator or by a device. A coordinator detects the presence of a PAN identifier conflict if the PAN coordinator subfield is set to 1 in the beacon received by the PAN coordinator. A device detects the presence of a PAN identifier conflict if a coordinator detects the presence of a PAN identifier conflict as long as the PAN coordinator subfield is set to 1 in the beacon received by the PAN coordinator and the address is not equal to macCoordShortAddress and macCoordExtendedAddress. This PAN identifier conflict is resolved by the coordinator by performing an active scan and establishing a new PAN identifier. A device resolves the PAN identifier conflict by sending a PAN identifier conflict notification to the coordinator. On receiving the notification the PAN coordinator sends an acknowledgment and resolves the conflict by performing an active scan and then selecting a new PAN identifier.

- *Starting a PAN:* After scanning all the channels and selecting a suitable PAN identifier, a PAN gets started by an FFD. A PAN is operated by an FFD by issuing the MLME-START.request primitive.

7.3.7 Association and Disassociation

After performing an active scan or passive scan, a device tries to associate with a coordinator by issuing the MLME-RESET.request primitive. A device can associate with a coordinator only if the macAssociationPermit of the coordinator is set to true. If the macAssociationPermit is set to false, then the association request from the device is ignored. A device gets associated with a coordinator that is willing to accept any associations. On receiving the association request command, the coordinator will send an acknowledgment frame to the device confirming that it has received the request from it for association.

The sending of an acknowledgment by the coordinator to the device itself does not mean that association is guaranteed. Based on the amount of resource available in the PAN, the coordinator decides whether or not the device should be associated. This decision is taken with an aResponseWaitTime symbols. If the device had associated with the PAN previously, then all the previous information related to that device will be removed. On availability of sufficient amount of resources, the device gets allocated with a short address by the PAN coordinator. Also an association response command is generated by the PAN coordinator to indicate a successful association status. If a sufficient amount of resources are not available, then the PAN coordinator generates an association response command with failure status indicating that the association has failed. The disassociation of a device from the PAN is carried out by the next upper layer by issuing the MLME-DISASSOCIATE.request primitive to the MLME. A disassociate request command is sent by the device to the PAN coordinator. The PAN coordinator then sends an acknowledgment frame indicating that it has successfully received the disassociation request and the device gets disassociated from the PAN.

7.3.8 Synchronization

Synchronization is achieved in a PAN with beacons and without beacons. In a beacon-enabled mode the device tries to synchronize with its PAN coordinator by issuing the MLME-SYNC.request primitive. When the device receives the beacon frame, it is checked to see whether the beacon frame it received has come from the coordinator to which it is associated. The source address and the source PAN identifier fields are checked to see if they match with the coordinator source address. If there is no match then the beacon is discarded. If there is a match, the beacon is declared as valid. Synchronization is achieved in a nonbeacon-enabled mode through polling.

7.4 Data Gathering Paradigm

The wireless sensor network is modeled as a directed graph $G = (V, E)$, where V is the set of nodes and E is the set of directed wireless links. Let S_S denote the set of sensor nodes and S_C denote the sink node or the coordinator node. Then, $V = S_S \cup S_C$. The transmission range for each sensor node is designated by r_{tx}.

Let d_{ij} denote the distance between node i and node j. A directed transmission link $(i, j) \in E$ exists if $d_{ij} \leq r_{tx}$.

All transmission links are assumed to be symmetrical, where $e_{ij} = e_{ji}$. A grid network topology as illustrated in Figure 7.6 (A 5 × 5 grid topology with 24 nodes and one base station) is considered for our study. In the grid topology pattern, each of the nodes can communicate with either horizontal nodes or the vertical nodes. All the nodes are static. The sensor network consists of beaconless coordinators with CSMA-CA mechanism. The CSMA-CA mechanism does not employ the RTS/CTS mechanism due to the low data rate nature of 802.15.4.

As described previously, communication takes place among nodes in horizontal and vertical direction as depicted in Figure 7.7. But, if a mobile sink is within the communication range of any node in any direction while it is moving, then communication takes place in that direction. Each of the nodes in the sensor network is beacon enabled with slotted CSMA-CA mechanism. Each of these coordinators is a fully functional routing device, allowing data transfer among each of the devices. The sensor nodes participate in the network throughout the simulation time. AODV is the underlying protocol for rout-

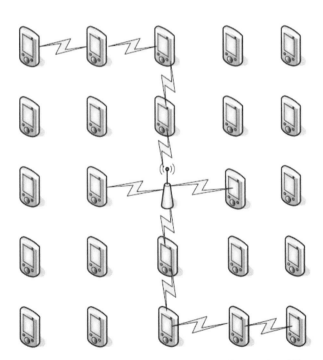

Figure 7.6 A 5×5 grid network topology with static nodes for results of Figure 7.8 through Figure 7.17 [29].

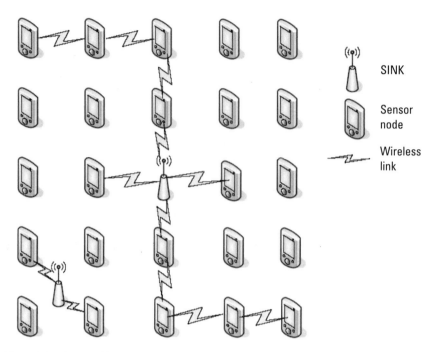

Figure 7.7 A 5×5 grid network topology with mobile ad hoc sink nodxe for results of Figure 7.18 through Figure 7.29 [30].

ing the data packets. Compeer communication takes place in sensor networks among various nodes for routing the data to the sink node.

7.5 Simulation Environment

Simulation is carried out for performance comparison, and we have modified the simulation code wherever it was deemed necessary to satisfy our simulation conditions and fine tuned several parameters to carry out our simulation work. The radio propagation model used is the two-ray ground model given by $P_r(d) = \frac{P_t G_r G_t h_r^2 h_t^2}{d^4 L}$, where h_t is the height of the transmitter antenna and h_r is the height of the receiver antenna. An omnidirectional antenna has been considered. The transmitter gain, receiver gain, and path loss values are considered 1.0. The simulations are conducted [25] using NS2 with WPAN extension by utilizing the standard specifications of Crossbow MICAz processor and radio platform (Chipcon CC2420), operating at the 2.4-GHz frequency band [26]. Here for the sensor network scenario, the data traffic is not generated in a many-to-many fashion. Instead, there is a designated sink node to communicate with the sensor nodes. This unique traffic pattern is modeled by modifying the cbrgen.tcl file, in which a node is designated as a sink node. This setting will result in the same simulation effect as suggested in [27].

The simulation results obtained for various performance metrics with traffic load and simulation time is based on the parameter mentioned in Table 7.1, while the results obtained for various metrics with different network scenarios is based on the simulation parameters mentioned in Table 7.2. We have selected various performance metrics such as PDR, average network delay, throughput of the network, normalized routing load, and dropped LQI packets for evaluating the effect of IEEE 802.15.4 over AODV for wireless sensor networks.

The normalized routing load is the number of routing packets "transmitted" per data packet "delivered" at the destination. It is the sum of all the control packets sent by all the sensor nodes in the network to discover and maintain routes to the SINK node. The dropped LQI packets is measured as the link quality indicator (LQI) that is considered to study the cause of packet drops and their effect on the performance of sensor network.

7.6 Result Analysis

Besides running independently, all the simulations are averaged for five different seeds. A static scenario is simulated for the sensor network, where all the sensor nodes have the same radio range. Energy is distributed uniformly among all the sensor nodes. Simulations are carried out in a beaconless mode (unslotted CSMA-CA). Each of these coordinator devices acts as fully functional devices.

Table 7.1
Simulation Parameters for Results from Figure 7.8–Figure 7.17

Parameter	Value
Routing protocol	AODV
MAC protocol	IEEE 802.15.4
Frequency/bandwidth	2.4 GHz/250 Kbps
Number of nodes	10, 15, 20, 25
Simulation area	50 x 50
Simulation time (sec)	50, 100, 150, 200
Queue Size	70
Traffic type	CBR
Packet size (bytes)	60
Traffic load (pps)	0.001, 0.01, 0.1, 0.3, 1.0, 5.0
Number of sources	4, 6, 8, 10, 12

Table 7.2
Simulation Parameters for Results
from Figure 7.18–Figure 7.29

Parameter	Value
Routing protocol	AODV
MAC protocol	IEEE 802.15.4
Frequency/bandwidth	2.4GHz/250kbps
BO	3
SO	3
Mobility model	Random trip mobility model
Sink speed	0m/s, 0.01m/s, 0.1m/s, 1m/s, 2m/s
Number of nodes	16
Simulation area	40 x 40
Simulation time (sec)	200
Queue size	70
Traffic type	CBR
Packet size (bytes)	60
Traffic load (pps)	0.001, 0.01, 0.1, 0.3, 1.0, 5.0
Number of sources	4, 6, 8, 10, 12
RxThresh	−94dBm
CSThresh	−94dBm

7.6.1 Evaluation of Various Metrics with Traffic Load

The increase of data rate is carried out starting from 0.001 packets per second to 5 packets per second. Figures 7.8 through 7.12 depict the effect of traffic on the performance of a sensor network. All the participating nodes have a PAN coordinator as the sink node. Collision in the network increases with the increase in traffic. For 10 and 15 nodes, the packet delivery ratio is close to 100 percent at 0.1 pkts/sec, while for 20 nodes the packet delivery is at 80

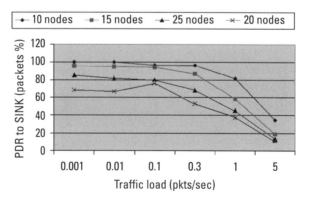

Figure 7.8 PDR to SINK vs. traffic load (pkts/sec) [29].

Performance Analysis of IEEE 802.15.4 with Sink Nodes 199

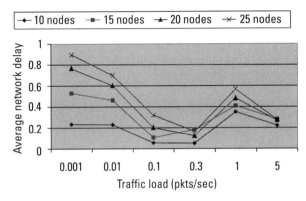

Figure 7.9 Average network delay vs. traffic load (pkts/sec) [29].

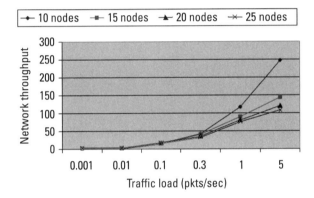

Figure 7.10 Network throughput vs. traffic load (pkts/sec) [29].

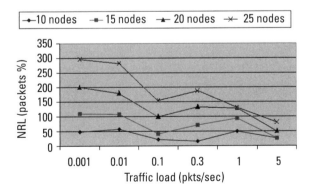

Figure 7.11 NRL (packets percent) vs. traffic load (pkts/sec) [29].

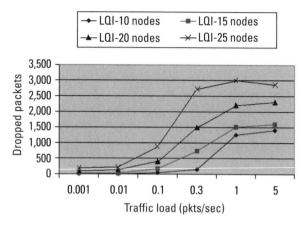

Figure 7.12 Dropped packets vs. traffic load (pkts/sec) [29].

percent (Figure 7.8). When the number of nodes is 25 in the network, the fall in packet delivery can be observed starting from 0.1 pkts/sec, indicating there is congestion in the network. There is a fall in packet delivery for all the node configurations once the traffic load is increased to above 0.3 pkts/sec. Dropping of packets in the network results in large number of retransmissions. This results in the huge overhead for 25 nodes (Figure 7.11).

There is a rise and fall in network delay between 0.9 to 0.6 sec. Networks with 25 nodes have higher delay. Some of the packets need to be transmitted by the intermediate nodes contributing to the delay (Figure 7.9). As the number of dropped packets increases in the network, there will automatically be a decrease in the number of packets delivered to the sink. Some nodes may remain out of the sensing range from other sensor source nodes. This results in a hidden node problem, where some nodes transmit packets without the knowledge that some other nodes might also be involved in the transmission. The throughput for 10 nodes is highest, followed by 15 nodes contour. It is observed that the throughput of the network increases as the number of packets increases (Figure 7.10). The throughput indicates the amount of data successfully received as the sink node. The link quality indication metric characterizes the strength and quality of the received packet. LQI measures the "incoming modulation of each successfully received packets" [28]. Figure 7.12 shows the dropped LQI packets. The highest drop of LQI packets is found for 25 nodes when compared to 10 and 15 nodes. Inefficiency in the backoff mechanism results in the drop of LQI packets.

7.6.2 Evaluation of Various Metrics with Simulation Time

Figures 7.13 through 7.17 depict the effects of the number of source nodes on the performance metrics. When the numbers of source nodes are 4 and

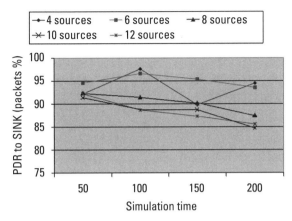

Figure 7.13 PDR to SINK vs. simulation time [29].

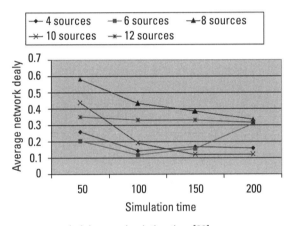

Figure 7.14 Average network delay vs. simulation time [29].

6, then the packet delivery is nearly 95 percent, which indicates that fewer source nodes contributes more packet delivery. As the number of source nodes increases, there is a steep fall in the amount of data delivered to the sink node. An increase in number of sources results in more routing traffic leading to less packet delivery. So there should be optimum number of sources. This confirms that 802.15.4 is meant for low traffic and simple applications.

7.6.3 Evaluation of Performance Metrics with Different Network Scenarios

We have considered different mobile speeds, 0 m/s, 0.01 m/s, 0.1 m/s, 1 m/s, and 2 m/s. Here, 0 m/s means that the sink node is not moving and the whole

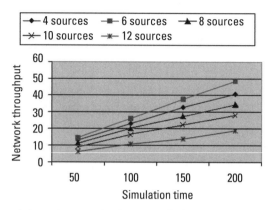

Figure 7.15 Network throughput vs. simulation time [29].

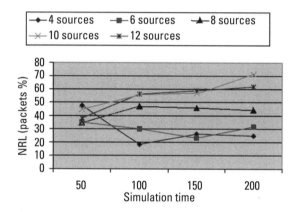

Figure 7.16 NRL (packets percent) vs. simulation time [29].

network is a static network. We wanted to check whether a mobile sink node results in any performance gain over a static sink node. In each of the simulations, the traffic load is varied from 0.001 pkts/sec to 5 pkts, the number of sources is varied from 4 to 12 on an increment of 2 sources at each stage, and the number of nodes is varied from 10 to 25 nodes. The number of sources is kept at 5, 8, 10, and 12 for 10, 15, 20, and 25 nodes, respectively.

7.6.3.1 PDR with Different Network Scenarios

Figures 7.18, 7.19, and 7.20 represent the amount of packets delivered to the sink at different speeds. It can be observed from the graph that less mobility contributes more packet delivery. When the sink node speed is less than 1 m/s, it is around the 70 to 98 percentile range. But once the number of packets is

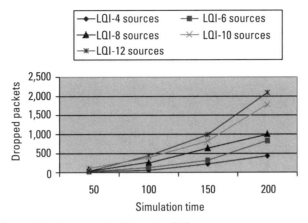

Figure 7.17 Dropped packets vs. simulation time [29].

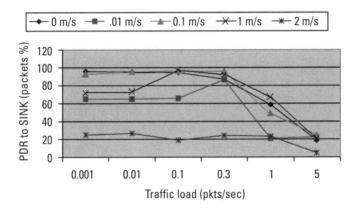

Figure 7.18 PDR to SINK vs. traffic load (pkts/sec) [30].

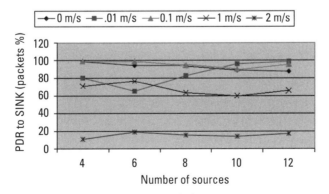

Figure 7.19 PDR to SINK vs. number of sources [30].

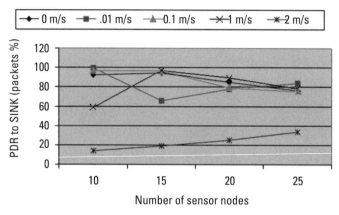

Figure 7.20 PDR to SINK vs. number of sensor nodes [30].

increased to more than 0.3 pkts/sec, a huge drop in performance is observed, no matter at what speed the sink node is moving. Delivery ratio decreases due to random backoffs and collisions. When compared to other sink mobile speeds, the worst performance is observed when the sink node moves at 2 m/s.

7.6.3.2 Average Network delay with Different Network Scenarios

Average network delay is marked by Figures 7.21, 7.22, and 7.23. It can be observed that a high speed of sink nodes does not guarantee that the packets are reached quickly at the sink node. Consider a situation where a node from one corner of the field needs to transmit the data to the sink node at the other end. The packet has to go through many intermediate nodes to reach the sink node. But if the sink node moves before receiving the packet, then the neighboring node has to again transmit to other intermediate nodes, which results in high end-to-end delay and high routing load in the network. Route discovery also

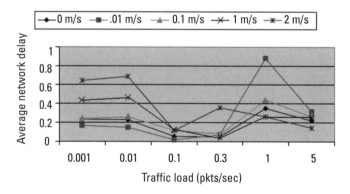

Figure 7.21 Average network delay vs. traffic load (pkts/sec) [30].

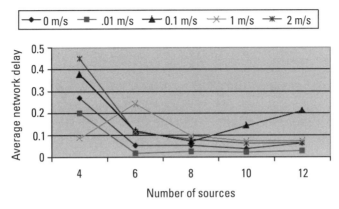

Figure 7.22 Average network delay vs. number of sources [30].

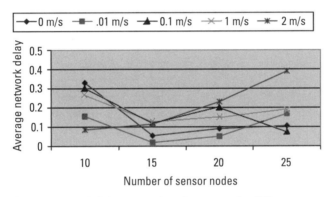

Figure 7.23 Average network delay vs. number of sensor nodes [30].

leads to delay. As can be observed from the graphs, the delay decreases when the packets and sources are varied. When the packets are increased to 0.5 pkts/sec and the sources to 6, then all the speeds converge indicating that the network is saturated.

7.6.3.3 Throughput with Different Network Scenarios

The throughput is drastically decreased when the sink node mobility speed is increased to 2 m/s, as can be seen from Figures 7.24 through 7.26. The throughput provides us with a surprise result. Throughput increases at a higher traffic rate due to a shorter contention window. The throughput of nodes moving at 0.1 m/s is high when compared to static scenario. This shows that higher throughput can be achieved with less mobility of sink nodes.

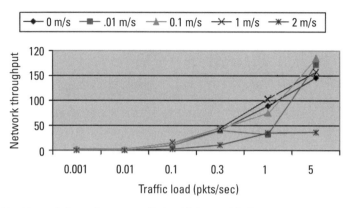

Figure 7.24 Network throughput vs. traffic load (pkts/sec) [30].

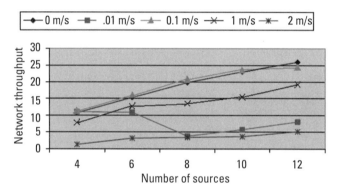

Figure 7.25 Network throughput vs. number of sources [30].

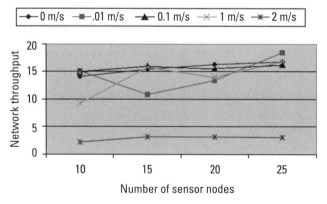

Figure 7.26 Network throughput vs. number of sensor nodes [30].

7.6.3.4 Normalized Route Load with Different Network Scenarios

When the nodes are static, and if each of the nodes is able to communicate with its neighboring node, then there will be fewer loads due to fewer packet requests for establishing the routes to the sink node and for association with the sink node. But if the sink is moving, then there can be association problems for the normal sensor nodes with the sink node. This results in more broken links, route requests, and route error messages to be generated in the network, as seen in Figures 7.27 through 7.29. This is more severe when the sink mobility speed is high or if it is more than 1 m/s.

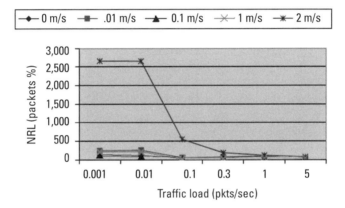

Figure 7.27 NRL (packets) vs. traffic load (pkts/sec) [30].

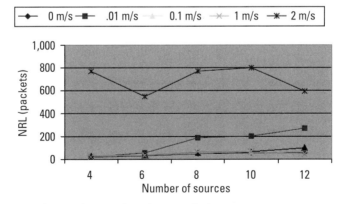

Figure 7.28 NRL (packets) vs. number of sources [30].

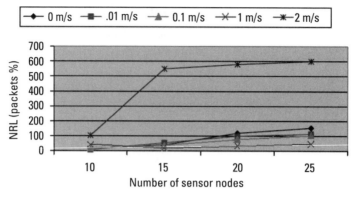

Figure 7.29 NRL (packets) vs. number of sensor nodes [30].

References

[1] Akylidiz, W., S. Sankarasubramaniam, and E.Cayrici, "A Survey on Sensor Networks," *IEEE Communications*, Vol. 40, No. 8, August 2002, pp.102–114.

[2] Akkaya, K., and M. Younis, "A Survey of Routing Protocols in Wireless Sensor Networks," *Elsevier Ad Hoc Network Journal*, 2005, pp. 325–349.

[3] Gowrishankar, S., T. G. Basavaraju, S. K. Sarkar, "Issues in Wireless Sensor Networks." In *Proceedings of the 2008 International Conference of Computer Science and Engineering, (ICCSE 2008)*, London, UK, July 2-4, 2008.

[4] Vincze, Z., et al., "Deploying Multicple Sinks in Multi-Hop Wireless Sensor Networks." In *Proceedings of the IEEE International Conference on Pervasive Services (ICPS)*, Istanbul, Turkey, July 15–20, 2007.

[5] Zheng, J., and M. J. Lee, "A Comprehensive Performance Study of IEEE 802.15.4," Chapter 4, in *Sensor Network Operations*, IEEE Press, Wiley InterScience, 2006, pp. 218–237. {AU: Is this book publised by IEEE Press or Wiley InterScience? What is the city of publication?}

[6] Yu, Q., J. Xing, and Y. Zhou, "Performance Research of the IEEE 802.15.4 Protocol in Wireless Sensor Networks." In *Proceedings of the 2nd IEEE/ASME International Conference on Mechatronic and Embedded Systems and Applications*, Beijing, January 29, 2007.

[7] Adams, L., "Capitalizing on 802.11 For Sensor Networks," Gainspan Corporation, Sunnyvale, CA.{AU: Is there a date?}

[8] Lye, P. G., and J. C. McEachen, "A Comparison of Optimized Link State Routing with Traditional Routing Protocols in Marine Wireless Ad-hoc and Sensor Networks," 40th Annual Hawaii International Conference on System Sciences (HICSS07), Waikoloa, HI, January 29, 2007.

[9] Lu, G., B. Krishnamachari, and C. S. Raghavendra, "Performance Evaluation of the IEEE 802.15.4 MAC for Low-Rate Wireless Network." In *Proceedings of the IEEE International Performance Computing and Communication Conference (IPCCC'04)*, Phoenix, AZ, April 2004, pp. 701–706.

[10] Gomez, C., et al., "Adapting AODV for IEEE 802.15.4 Mesh Sensor Networks: Theoretical Discussion and Performance Evaluation in a real Environment." In *Proceedings of the IEEE 2006International Symposium on a World of Wireless, Mobile and Multimedia Networks (WoWMoM 06)*, Buffalo-Niagara Falls, NY, July 2006.

[11] Hoffert, J., K. Klues, and O. Orjih, "Configuring the IEEE 802.15.4 MAC Layer for Single Sink Wireless Sensor Network Applications," Real Time Systems Class Project, Washington University, St. Louis, MO, December 2005.

[12] Lee, J.-S., "An Experiment on Performance Study of IEEE 802.15.4 Wireless Networks," 10th IEEE Conference on Emerging Technologies and Factory Automation (ETFA05), Catania, Italy, September 19–22, 2005.

[13] Athanasopoulos, A., E. Topalis, C. Antonopoulos, S. Koubias, "802.15.4: The Effect of Different Back-Off Schemes on Power and QOS Characteristics," Third International Conference on Wireless and Mobile Communications (ICWMC'07), Guadeloupe, Mexico, March 4–9, 2007.

[14] Ko, J.-G., Y.-H. Cho, and H. Kim, "Performance Evaluation of IEEE 802.15.4 MAC with Different Backoff Ranges in Wireless Sensor Networks," 10th IEEE Singapore International Conference on Communication Systems (ICCS06), Singapore, October 2006.

[15] Chen, C., and J. Ma, "Simulation Study of AODV Performance over IEEE 802.15.4 MAC in WSN with Mobile Sinks," 21st International Conference on Advanced Information Networking and Applications Workshops (AINAW'07), Niagara Falls, Ontario, Canada, May 21–23, 2007.

[16] IEEE Standard for Part 15.4: Wireless Medium Access Control Layer (MAC) and Physical Layer (PHY) Specifications for Low Rate Wireless Personal Area Networks (LR-WPANs), IEEE Std 802.15.4, 2003.

[17] Guteirrez, J. A., et al., "Low Rate Wireless Personal Area Networks: Enabling Wireless Sensors with IEEE 802.15.4," *IEEE Press*, November 2003.

[18] Zheng, J., and M. J. Lee, "Will IEEE 802.15.4 Make Ubiquitous Networking a Reality? A Discussion on a Potential Low Power Low Bit Rate Standard," *IEEE Communications*, June 2004, pp. 140–146.

[19] Rao, V. P., "The Simulative Investigation of Zigbee/IEEE 802.15.4," Master's Thesis, Department of Electrical and Information Technology, Dresden University of Technology.

[20] Grammer, J., "Zigbee Starts to Buzz," *IEEE Review*, November 2004.

[21] Ramachandran, I., A. K. Das, and S. Roy, "Analysis of the Contention Access Period of IEEE 802.15.4 MAC," *ACM Transactions on Sensor Network*, Vol. 3, No. 1, Article No. 4, March 2007.

[22] Naeve, M., "IEEE 802.15.4 Overview," IEEE 802.15.4a Task Group, 2004.

[23] Bougard, B., F. Catthoor, D. Daly, D. C. Chandrakasan, and A. Dehaene, "Energy Efficiency of the IEEE 802.15.4 Standard in Dense Wireless Microsensor Networks: Modeling and Improvement Perspectives," Proceedings of Design, Automation and Test in Europe, March 7–11, 2005, Vol. 1., pp. 196–201.

[24] Rao, V. P., and D. Marandin, "Adaptive Channel Mechanism for Zigbee (IEEE 802.15.4)," *Journal of Communications Software and Systems*, Vol. 2, No. 4, December 2006, pp. 283–293.

[25] "Zigbee Software Modules," City College of New York, http://ees2cy.engr.ccny.cuny.edu/zheng/pub, las accessed 2010.

[26] http://www.xbow.com/Products/Product_pdf_files/Wireless_pdf/MICAZ_Datasheet.pdf, last accessed 2009.

[27] Downard, I., "NRL's Sensor Network Extension to NS-2," NRL/FR/04-10073, Naval Research Laboratory, Washington, DC, May 2004, http://www.nrlsensorsim.pf.itd.nrl.navy.mil, last accessed 2009.

[28] Wang, Y., et al., "A Supervised Learning Approach for Routing Optimizations in Wireless Sensor Networks." In *Proceedings of the 2nd International Workshop on Multihop Ad Hoc Networks: From Theory to Reality*, Florence, Italy, 2006.

[29] International Conference on Recent Advances in e-Commerce and i-Technologies, Chennai, India.

[30] International World Congress on Engineering and Computer Science 2010, IAENG, USA.

8

Performance Evaluation and Traffic Load Effect on Patrimonial ZigBee Routing Protocols in WSNs

8.1 Introduction

The performance of sensor networks is closely related to the rate at which the source nodes generate the traffic. Care should be taken that the generated traffic does not congest and degrade the performance of the network. Thus, it is necessary to model the network's traffic characteristics accurately in order to develop efficient protocols for wireless sensor networks.

In this chapter, we study the exponential on/off traffic effect on the performance of wireless sensor networks using IEEE 802.15.4/ZigBee standard. The simulation is performed using NS2 with WPAN extension, which is an open source network simulator tool. Performance discrepancies in sensor networks under various traffic scenarios are still largely obscure. There is a need to understand the effect of traffic on versatile behavioral aspects of wireless sensor networks like number of sources and sensor nodes. Although this aspect has been studied in ad hoc networks, it is still largely an unexplored area in sensor networks. Through simulation, it has been shown that the exponential on/off traffic of ZigBee routing scheme with IEEE 802.15.4 MAC association offers several benefits for a typical WSN application.

The study of traffic load effect on the performance of wireless sensor networks was inspired from a research paper by Au-Yong et al. [1]. He studied the exponential on/off traffic effect on the performance of on-demand wireless mobile ad hoc routing protocols. In our work, we have modeled the on/off traffic for a wireless sensor network scenario and have studied it from the same point of view.

The main contributions of this chapter are as follows:

- This study is the first of its kind to compare the performance of AODV family of routing protocols, namely, AODV, AODVUU, AOMDV, and RAODV, from a sensor network point of view. To the best of our knowledge, no work has been reported that compares and studies the performance of all these routing protocols for wireless sensor networks using IEEE 802.15.4 as the underlying MAC layer.
- Substantial amount of effort has been made to study the effect of exponential on/off traffic effect on the performance of wireless sensor networks using IEEE 802.15.4/ZigBee standard. To the best of our knowledge, an insubstantial amount of work has been reported for studying the traffic effect on the performance of wireless sensor networks using the ZigBee/IEEE 802.15.4 standard.

8.2 Literature Background

A performance comparison of AODV and DSR is done under a mobile ad hoc scenario by Au-Yong [1]. Here, comparison of both the protocols is done under the effect of on/off source traffic with exponential distribution in comparison to constant bit rate traffic. The authors show how the protocol's behavior changes unexpectedly when it is put under on/off traffic source instead of the CBR model. DSR outperforms AODV under the on/off traffic scenario.

Performance comparison of two routing protocols (AODV and AOMDV) is done by Marina et al. [2] by varying the node mobility and the traffic load in a mobile ad hoc environment. When the mobility is increased, the packet delivery ratio of both AODV and AOMDV decreases, and it is more severe in AODV. The delay in AOMDV is substantially reduced due to the availability of alternate routes. When the number of packets is generated at a very low rate, then both AODV and AOMDV behave in the same manner, but when the traffic load is increased, AOMDV outperforms AODV, as it can take care of link breakages at high traffic rate.

Reverse AODV (RAODV) and AODV is compared by Kim et al. [3]. Various metrics like delivery ratio, average end-to-end delay, average energy remained, and control overheads are considered. Here the average energy metric is the energy remained in each and every node. The authors through simulation show that RAODV nodes have more energy left as compared to AODV. RAODV has better performance when delivering the packets, but it has high control packet overhead as it floods the network with more reverse route reply messages.

Performance of LEACH protocol is analyzed over IEEE 802.15.4. The original LEACH protocol was modified to select a cluster head. There is a clear distinction between FFD and RFD devices. The parameters considered for evaluation are (a) total number of packets received at the BS, (b) the number of nodes that are alive at the end of the simulation, and (c) the longevity of the network. The topological scenarios considered are simple LEACH with homogeneous network and hierarchical clustering with homogeneous and heterogeneous network. It is shown that for the amount of messages received at the BS, the homogeneous single hop network has good performance, while heterogeneous network has worst performance. The amount of energy consumed is more for heterogeneous topology. It is also observed that the RFD nodes side more quickly in the heterogeneous network than in the homogeneous network [4].

The performance comparison of three routing protocols is done in a self-similarity traffic condition [5]. The protocols compared are AODV, DSR, and OLSR. The traffic models considered are CBR, Pareto, and exponential. Various parameters like packet delivery ratio, end-to-end delay, normalized routing overhead, and throughput are mapped against mobility speed. AODV and OLSR have high packet delivery rates compared to DSR. OLSR has the highest overhead due to its proactive nature, while DSR has highest throughput. AODV has less delay compared to other routing protocols.

In Goh et al. [6], two types of packets are considered. They are priority packets and normal packets. Some nodes that are near to the phenomenon are considered important, and they are designated *priority nodes*. The traffic generated from these nodes is called *priority traffic*. Traffic from the remaining nodes is called *normal traffic*. The performance of priority packets was evaluated by considering the single hop mode and multihop mode. Through simulation it is shown that the priority packets have high throughput and low delay in single hop and multihop communication.

A performance comparison of SMAC and a traffic-aware energy-efficient MAC protocol is carriedout in Suh et al.[7]. Both the protocols were evaluated using MICA mote. The proposed MAC protocol is adaptive to wake and sleep state by piggybacking on the traffic generated by the nodes. It is shown that the proposed MAC protocol consumes less energy SMAC.

REALMOBGEN software was developed by Doci et al. [8]. The developers argue that mobility and traffic are interconnected and they have the REALMOBGEN software to generate scenarios based on the mentioned argument. Using the tool, two protocols (AODV and DSR) are compared. These two routing protocols were evaluated by varying the speed, pause time, number of sources, and rate of packets generated. The authors have proposed a new parameter called *availability*, which is used to evaluate the scenarios generated by the REALMOBGEN tool.

8.3 Patrimonial ZigBee Routing Protocols

IEEE 802.15.4 and ZigBee are industry standards designed to be used in low-data-rate, low-power-consumption, low-cost, and long-lived networks. IEEE 802.15.4 is sometimes called ZigBee, even though ZigBee specifically refers to the routing protocol and 802.15.4 refers to the MAC and PHY protocols. The ad-hoc on demand distance vector (AODV) routing protocol has been recommended to be used as the routing protocol for ZigBee [9, 10].

Patrimonial is defined as inherited or inheritable by established rules (usually legal rules) of descent. The routing protocols AODVUU, AOMDV, and RAODV are built by inheriting the features of AODV routing protocol. We argue that these routing protocols can be called the AODV family of protocols, as all these protocols consider AODV the base routing protocol upon which these protocols are developed by incorporating some special features. Since AODV is the proposed routing protocol for the ZigBee standard, these AODVUU, AOMDV, and RAODV routing protocols can be considered patrimonial ZigBee routing protocols.

In this chapter we focus on the performance study of four routing protocols—AODV, AODVUU, RAODV, and AOMDV. Even though AODV and AODVUU are not different protocols, we want to see if there is any improvement in using the AODVUU implementation for a sensor network environment. We have investigated whether a multiple path algorithm like AOMDV would result in more data delivery as compared to single path solutions like AODV in a sensor network. Also, the reverse route discovery mechanisms employed in RAODV is checked for a sensor network. There is a need to understand the versatile behavioral aspects of these routing protocols in a wireless sensor network with varying traffic loads and the number of sources.

8.4 Traffic Generators

In NS2 four different types of on-off models are supported [11]. They are exponential, Pareto, CBR, and Poisson.

8.4.1 Exponential On-Off Traffic

Generation of packets is done at a fixed rate during on period while no packets are generated during off period [12]. The following settings are involved for the generation of exponential on-off traffic.

```
set expotraffic [new Application/Traffic/Exponential]
$expotraffic set PacketSize_
$expotraffic set rate_
$expotraffic set burst_time_
$expotraffic set idle_time_
```

8.4.2 Pareto On-Off Traffic

While generating Pareto on-off traffic, its shape value has to be specified [13]. The following settings are involved for the generation of Pareto on-off traffic.

```
set paretotraffic [new Application/Traffic/Pareto]
$paretotraffic set PacketSize_
$paretotraffic set rate_
$paretotraffic set burst_time_
$paretotraffic set idle_time_
$paretotraffic set shape_
```

8.4.3 CBR On-Off Traffic

In CBR traffic the packets are generated at a constant rate. We need to specify the starting time and stopping time of the traffic [14, 15]. The following settings are involved for the generation of CBR on-off traffic.

set cbrtraffic [new Application/Traffic/CBR]
$cbrtraffic set PacketSize_
$cbrtraffic set rate_
$cbrtraffic set interval_

8.4.4 Poisson On-Off Traffic

Exponential on-off Poisson traffic can be generated either by specifying mean time between the packets or long time sending rate [16]. The following settings are involved for the generation of Poisson on-off traffic.

```
set poissontraffic [new Application/Traffic/Poisson]
with mean time
$poissontraffic set interval_
$poissontraffic set packetSize_
$poissontraffic set maxpkts_
With long term sending term
$poissontraffic set rate_
$poissontraffic set packetSize_
$poissontraffic set maxpkts_
```

8.5 Traffic Model

Self-similarity is an essential characteristic for traffic in communication network traffic. We would like to quote the self-similarity process as in [17, 18]. A stochastic covariance stationary is represented as $X = (X_j : j = 0, 1, 2...)$. When an original X series is averaged over nonoverlapping blocks, then it results in covariance stationary time. A covariance stationary time can be represented as

$$X^{(m)} = \left(X_k^{(m)} : k = 1, 2, 3, ...\right)$$

where $m = 1, 2, 3, ...$

$$X_k^{(m)} = \frac{1}{m}\left(X_{km-m+1} + \cdots + X_{km}\right), k \geq 1$$

The process X is self similar with the self similarity parameter $H = 1 - \beta/2$.

The *on* period is defined as the time interval during which the packets are generated uniquely. The *off* time is defined as any time interval where the packets are not generated. Traffic is generated in burst. The exponential *on/off* traffic model consists of source nodes transmitting packets alternatively between *on* and *off* periods. As specified in [19], there is no need to cohere to this model (i.e., sometimes the *on* period can be followed by an *on* period, and an *off* period can be followed by another *off* period).

8.6 Simulation Environment

The wireless sensor network is modeled as a directed graph $G = (V, E)$, where V is the set of nodes and E is the set of directed wireless links. Let S_C denote the set of sensor nodes and S_P denotes the sink node.

Then, $V = S_C \cup S_P$. All sensor nodes have a fixed transmission range of r_{tx}. Let d_{ij} denote the distance between node i and node j. A directed transmission link (i,j) CE exists if $d_{ij} \leq r_{tx}$.

All transmission links are assumed to be symmetrical, where $e_{ij} = e_{ji}$.

We have modified the simulation code as explained in Section 7.5 [20]. Here, for the sensor network scenario the data traffic is not generated in an N-by-N fashion. Instead, there is a designated sink node to communicate with the sensor nodes. This unique traffic pattern is modeled by modifying the cbrgen.tcl file, in which a node is designated as a sink node, thereby enabling to achieve a similar effect as suggested in [21]. The results obtained for simulating various metrics with different network scenarios for AODV protocol is as per the simulation parameters mentioned in Table 8.1, while the results obtained for

Table 8.1
Various Parameter for Simulation of AODV

Parameter	Value
Routing protocols	AODV
MAC protocol	IEEE 802.15.4
Frequency/bandwidth	2.4 GHz/250 Kbps
BO	3
SO	3
Number of nodes	10, 15, 20, 25
Simulation area	40 x 40
Simulation time (sec)	200
Queue size	70
Packet size (bytes)	60
Traffic load (pkts/sec)	0.01, 0.1, 0.3, 0.4, 0.5, 1.0, 3.0, 5.0
Number of sources	4, 6, 8, 10, 12
RxThresh	−94 dBm
CsThresh	−94 dBm

simulating various metrics with different network scenarios for AODV, AODVUU, RAODV, and AOMDV protocols are based on the simulation parameters depicted in Table 8.2. We have selected metrics like PDR, average network delay, throughput, NRL, and power consumed for evaluating the effect of IEEE 802.15.4 over AODV for wireless sensor networks. All simulations are run independently and their results averaged at five different seeds. The on duration is set at 5 sec, and off duration is set at 2 sec. For our simulation we have assumed that the sensor network is static, where all the sensor nodes have the same radio range and energy is uniformly distributed among all the sensor nodes.

Table 8.2
Various Parameters Used for Simulation of Various Protocols

Parameter	Value
Routing protocols	AODV, AODVUU, RAODV, AOMDV
MAC protocol	IEEE 802.15.4
Frequency/bandwidth	2.4GHz/250kbps
Number of nodes	16
Simulation area	40 x 40
Simulation time (sec)	200
Queue size	70
Packet size (bytes)	60
Traffic load (pkts/sec)	0.001, 0.01, 0.1, 0.3, 1.0, 3.0, 5.0
Number of sources	4, 6, 8, 10, 12

8.7 Result Analysis

The simulations are carried out in a beacon-enabled mode, and all the devices have the capabilities of a coordinator (FFD) (i.e., to handle association and relay data packets).

8.7.1 Performance Evaluation of Various Protocols Under Network Scenarios

Figures 8.1 through 8.3 indicate the packet delivery ratio at various scenarios. The comparison of the impact of the two traffic schemes on the PDR is given

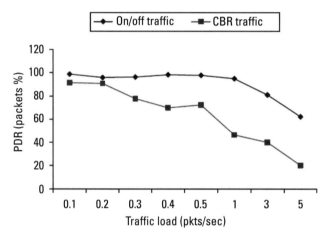

Figure 8.1 PDR vs. traffic load for traffic scenarios [22].

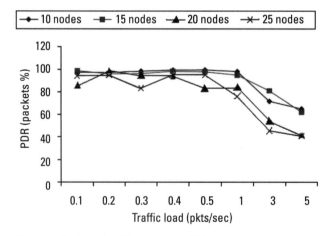

Figure 8.2 PDR vs. traffic load for different nodes [22].

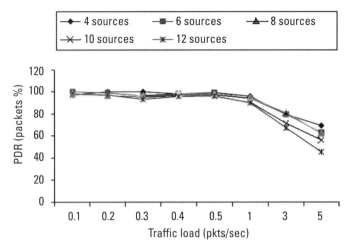

Figure 8.3 PDR vs. traffic load for different sources [22].

in Figure 8.1. The exponential on/off traffic has better performance when compared to the CBR traffic performance, as can be seen from the graph. The exponential on/off traffic is in the 90th percentile range up to 1 pkts/sec, while the CBR traffic losses its performance at the 0.3 pkts/sec. This clearly describes the on/off traffic performance over CBR traffic performance. The PAN devices have to periodically track the super frame beacon in order to synchronize to the PAN coordinator and get the parameters such as SO, BO, and CAP. Fewer transmission collisions occur in the CAP. The packet delivery decreases due to higher number of neighboring nodes and higher packet collision probability.

Average network delay is depicted by Figures 8.4 through 8.6. Some of the packets are lost due to collisions, resulting frequent retransmissions. This not only increases the delay due to frequent retransmissions, but also increases the size of the backoff windows, thereby negatively influencing the network throughput. Route discovery phase results in packet loss, and flooding of route discovery messages increases the delay. Data packet collisions may cause routing paths to fail and the route discovery mechanism will be triggered again. Collisions caused by route failures coupled with retransmissions at the 802.15.4 MAC layer, results in higher delays in the network.

As seen from Figures 8.7 through 8.9, the exponential on/off traffic produces better performance throughput. In general, we see that as the number of packet bits increases, the throughput efficiency also increases. Throughput is more in the exponential on/off scheme when compared with the CBR traffic scheme.

Figures 8.10 through 8.12 represent the normalized routing load. NRL in the CBR traffic is more when compared to on/off traffic. Eventually there is an increase in both CBR traffic and the exponential on/off traffic as the number

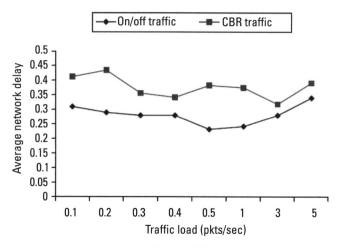

Figure 8.4 Average network delay vs. traffic load [22].

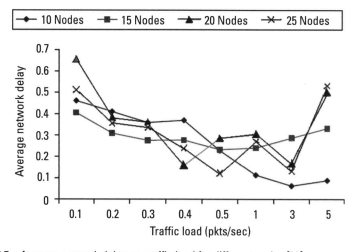

Figure 8.5 Average network delay vs. traffic load for different nodes [22].

of packets generated is increased. As can be seen from Figure 8.9, initially when the packets generated is less than 0.2 pkts/sec then the sensor network with 20 and 25 node configuration has less NRL, but as and when the packets are increased then the NRL increases for network with higher node configurations. The best NRL performance can be obtained when the number of sources is less than six.

Having the same initial energy across all the nodes in the network helped us to understand the system performance on the basis of energy. This can be made out through Figures 8.13 through 8.15. The average energy consumption

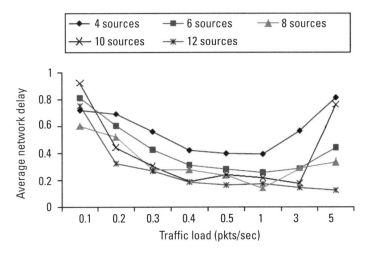

Figure 8.6 Average network delay vs. traffic load for different sources [22].

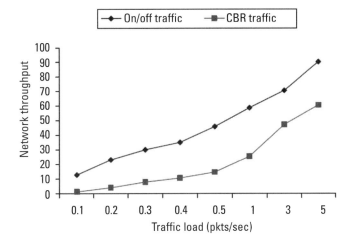

Figure 8.7 Network throughput vs. traffic load [22].

increases as the traffic is varied from 0.1 pkts/sec to 5 pkts/sec. But the energy consumption is more in case of CBR traffic when compared to the exponential on/off traffic. Synchronization of channel in the IEEE 802.15.4 MAC layer is mainly done by keeping track of the superframe beacons emitted from the PAN coordinator, which is also an overhead contributing to lower power efficiency. Substantial amount of energy is consumed by periodical transmission and tracking of the beacons.

The IEEE 802.15.4 standard is designed for low data rate applications. So the traffic load is varied from 0.001 pkts/sec to 5 pkts/sec. The number of

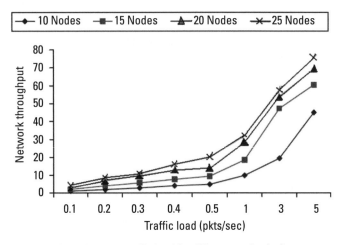

Figure 8.8 Network throughput vs. traffic load for different nodes [22].

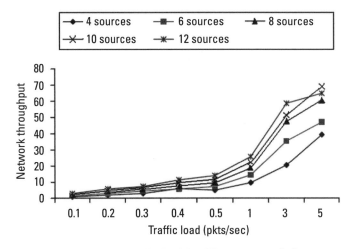

Figure 8.9 Network throughput vs. traffic load for different souces [22].

sources that generated the packets is varied starting from 4 and incremented on a scale of 2 up to 12 sources.

PDR of the various routing protocols is represented by Figure 8.16 and Figure 8.20. We observe that AODV, AODVUU, and AOMDV protocols remain in the 90 percentile range for up to 3 pkts/sec but RAODV has the worst performance, with it being in the 70 percentile range and dropping off very significantly as the packets are varied. Among all the four routing protocols, RAODV has the worst performance when the traffic load is varied.

Performance Evaluation and Traffic Load Effect 223

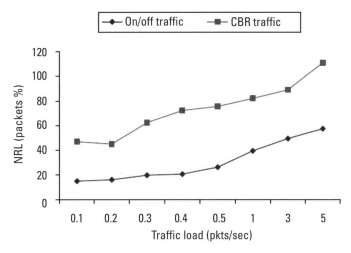

Figure 8.10 NRL vs. traffic load for different traffic [22].

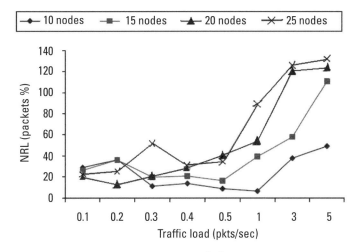

Figure 8.11 NRL vs. traffic load for different nodes [22].

The performance of AODV, AODVUU, and AOMDV is more or less the same for the varying traffic load before dropping off significantly. Due to multipath in AOMDV, there can be many stale routes which may contribute to less packet delivery and increase of routing overhead in the network. This shows that IEEE 802.15.4 is mainly for low-data-rate applications. In RAODV the packets are dropped due to the collision from additional reverse route reply packets sent during route discovery. But the trend is reversed between AOM-DV and RAODV when the number of sources is varied. Even though both

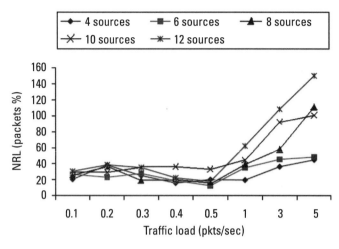

Figure 8.12 NRL vs. traffic load for different sources [22].

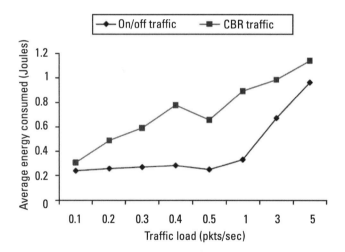

Figure 8.13 Average energy consumed vs. traffic load for different traffic [22].

AOMDV and RAODV both suffer from significant drop in packet delivery, it is more severe in AOMDV when compared with other RAODV, which was unexpected. As usual AODV and AODVUU maintain their performance in the 90th percentile range.

Even though AODV and AODVUU are one and the same routing protocols, they are implemented by different groups. AODV is the default implementation that can be found in NS2 tool, while AODVUU is the implementation of AODV routing protocol by Uppsala University. Both AODV and

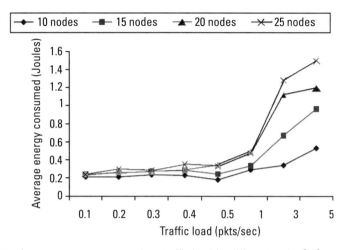

Figure 8.14 Average energy consumed vs. traffic load for different nodes [22].

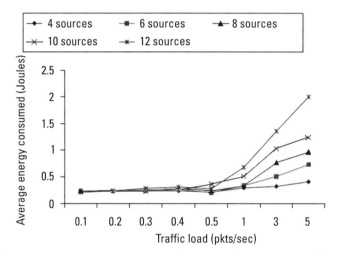

Figure 8.15 Average energy consumed vs. traffic load for different sources [22].

AODVUU have comparable PDR, but in terms of average network delay there is a huge difference between the performance of AODV and AODVUU. AODVUU has less network delay when compared with AODV. The delay of the AODV, RAODV, and AOMDV routing protocols decreases and converges at a point as the packets are varied, indicating that the network gets saturated as seen in Figures 8.18 and 8.19. Here in AOMDV, the duplicate copies are not discarded immediately, as they are used for further route discovery. This leads to more end-to-end delay in the network.

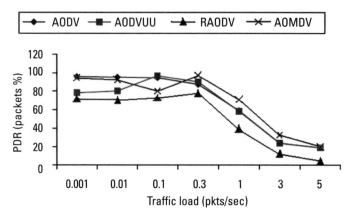

Figure 8.16 PDR vs. traffic load (pks/sec) for different routing protocols [23].

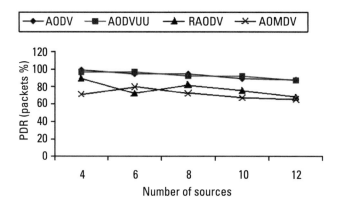

Figure 8.17 PDR vs. number of sources for different rating protocols [23].

The throughput of various routing protocols can be seen from Figures 8.20 and 8.21. When the routing load is varied, AOMDV has the highest throughput, while AODV is having the highest throughput when the number of sources is varied. This shows that even though AOMDV has the highest throughput, it does not guarantee the delivery of packets in a less constrained environment like a sensor network. AODV maintains a steady throughput while varying the traffic source and traffic load.

The routing load of AOMDV and RAODV is more when compared to the AODV routing protocol, as shown in Figures 8.22 and 8.23. The size of the control packets is high in AOMDV and RAODV due to extra route discovery mechanisms. This arises due to the various routing mechanisms incorporated into these routing protocols. In AOMDV, multipath allows the packets to move

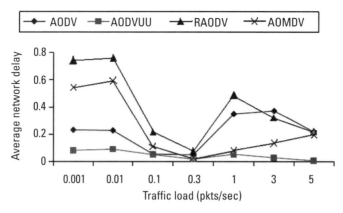

Figure 8.18 Average network delay vs. traffic load (pks/sec) for different routing protocols [23].

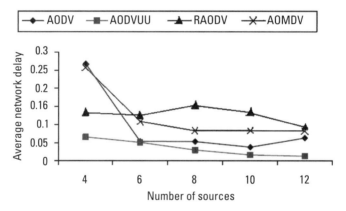

Figure 8.19 Average network delay vs. number of sources for different routing protocols [23].

in many paths, thus increasing the frequency of route reply. RAODV, on the other hand, uses reverse path technique to find the paths that naturally increases the number of control packets needed to keep the track of the increasing number of paths. Here for the simulation, since it is assumed that the nodes are static, link failures is very rare and hence computing for link failures will lead to additional overhead in AOMDV, as evidenced from Figures 8.22 and 8.23. Also RAODV floods the network with a huge amount of reverse route reply packets for route discovery from the sink to the other sensor nodes, which is unnecessary in the present setup as all the nodes have equal energy and the same transmission range, and they participate throughout the simulation.

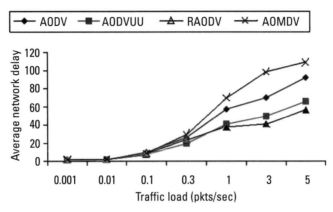

Figure 8.20 Network throughput vs. traffic load (pks/sec) for different routing protocols [23].

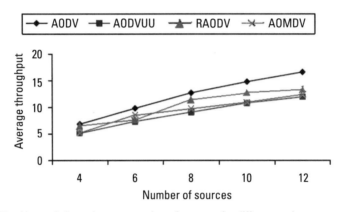

Figure 8.21 Network throughput vs. number of sources for different rating protocols [23].

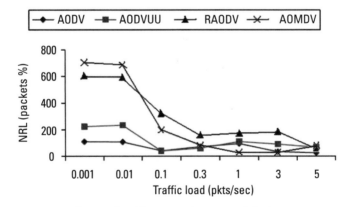

Figure 8.22 NRL vs. traffic load for different routing protocols [23].

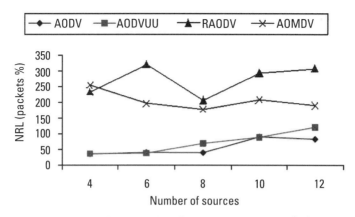

Figure 8.23 NRL vs. number of sources for different routing protocols [23].

References

[1] Au-Yong, J.-H., "Comparison of On-Demand Mobile Ad Hoc Network Routing Protocols Under On/Off Source Traffic Effect." In *Proceedings of the IASTED International Conference Networks and Communication Systems*, Chiang Mai, Thailand, March 29–31, 2006.

[2] Marina, M. K., and S. R. Das, "On-Demand Multipath Distance Vector Routing in Ad-Hoc Networks." In *Proceedings of the IEEE International Conference on Network Protocols*, 2001, pp. 14–23.

[3] Kim, C., E. Talipov, and B. Ahn, "A Reverse AODV Routing Protocol in Ad Hoc Mobile Networks," LNCS 4097, EUC Workshops, 2006, pp. 522–531.

[4] Gupta, A., A. Delye, et al., "Effect of Topology on the Performance of Mobile Heterogeneous Sensor Networks." In *Proceedings of Med Hoc Net*, Ionian Academy, Corfu, Greece, June 12–15, 2007.

[5] Maashri, A., and O. Khaoua, "Performance Analysis of MANET Routing Protocols in the Presence of Self Similar Traffic." In *Proceedings of the 31st IEEE Conference on Local Computer Networks*, Tampa, FL, November 14–16, 2006.

[6] Goh, H. G., et al., "Performance Evaluation of Priority Packet for Wireless Sensor Network," 2008 Second International Conference on Sensor Technologies and Applications, 2008.

[7] Suh, C., and Y.-B. Ko, "A Traffic Aware, Energy Efficient MAC Protocol for Wireless Sensor Networks." In *Proceedings of IEEE 2005 International Symposium on Circuits and Systems*, New York, May 2005.

[8] Doci, A., "Interconnected Traffic with Real Mobility Tool for Ad hoc Networks." In *Proceedings of 2008 International Conference on Parallel Processing*, 2008.

[9] Hoffert, J., K. Klues, and Obi Orjih, "Configuring the IEEE 802.15.4 MAC Layer for Single Sink Wireless Sensor Network Applications," Real Time Systems Class Project, Washington University, St. Louis, MO, December 2005.

[10] IEEE Std 802.15.4, IEEE Standard for Information Technology, Telecommunications, and Information Exchange Between Systems, Local and Metropolitan Area Networks, Part 15.4: Wireless Medium Access Control (MAC) and Physical Layer (PHY) Specifications for Low-Rate Wireless Personal Area Networks, IEEE Computer Society.

[11] http://www.isi.edu/nsnam/ns/doc-stable/ns_doc.pdf.

[12] Figueiredo, D. R., and B. Liu, "On the Specification of NS and Other Known On-Off Sources," Computer Science Technical Report 00-25, Department of Computer Science, University of Massachussetts, Amherst, MA, June 2000.

[13] http://nile.wpi.edu/NS/Howto/App_pareto_on_off_html, last accessed 2009–2010.

[14] http://www.it.usyd/edu.au/~deveritt/networksimulation/trafficgeneration.html, last accessed 2009–2010.

[15] http://amitkeerti.blogspot.com/2009/08/ns2-simulation-traffic-generators.html, last accessed 2009–2010.

[16] Pentikousis, K., "Application/Traffic/Poisson: Poisson Traffic Generator for ns-2," NS-2 Module, September 2004, http://ipv6.willab.fi/kostas/src/Application-Traffic-Poisson/, last accessed 2009–2010.

[17] Willinger, W., et al., "Self-Similarities in High Speed Packet Traffic: Analysis and Modeling of Ethernet Traffic Measurements," *Statistical Science*, Vol. 10, No. 1, 1995, pp. 67–85.

[18] Leland, W. E., M. S. Taqqu, W. Willinger, and D. V. Wilson, "On the Self-Similar Nature of Ethernet Traffic (Extended Verison)," *IEEE/ACM Transactions on Networking*, Vol. 2, No. 1, February 1994.

[19] Yin, S., and X. Lin, "Traffic Self Similarity in Mobile Ad Hoc Networks," Department of Electronic Engineering, Tsinghua University, China, IEEE, 2005.

[20] Downard, I., "NRL's Sensor Network Extension to NS-2," NRL/FR/04-10073, Naval Research Laboratory, Washington, DC, May 2004, http://www.nrlsensorsim.pf.itd.nrl.navy.mil/, last accessed 2009–2010.

[21] http://www.xbow.com/Products/Product_pdf_files/Wireless_pdf/MICAZ_Datasheet.pdf, last accessed 2010.

[22] International Conference on Sensors and Related Technologies (SENNET09), Chennai, India.

[23] IEEE International Conference on Education, Information Technology Applications (EITA 2009), Beijing, China.

9

Applications and Recent Developments

9.1 Introduction

Wireless mobile ad hoc networks and sensor networks have experienced a rapid expansion due to proliferation of inexpensive, smaller, and widely available more powerful mobile devices [1–6]. They are gaining importance with an increasing number of widespread applications because of the increase of portable devices as well as progress in wireless communication, which includes data rates compatible with multimedia applications, global roaming capability, and coordination with other network structures [7–11]. MANET has been established as an essential part of a future pervasive computing environment because of its intrinsic flexibility, lack of infrastructure, auto-configuration, ease of deployment, low cost, and potential applications. Because of the omnipresent existence of portable wireless computers, MANETs are increasingly used in search and rescue operations, disaster management, home automation, the private sector, the commercial sector, and the defense sector, providing commuters/users to access and information exchange regardless of their geographic situation or proximity infrastructure [12–14].

A wireless sensor network (WSN) is a special type of ad hoc network. Recent advances in wireless and electronic technologies have enabled a wide range of applications of WSNs in military, traffic surveillance, target tracking, environment and habitat monitoring, healthcare monitoring, industrial process monitoring and control, machine health monitoring, home automation, and so on [15–20]. Many new challenges have surfaced for the designers of WSNs, in order to meet the requirements of various applications like sensed quantities, size of nodes, and nodes' autonomy [21–25]. Therefore, improvements in the current technologies and better solutions to these challenges are required. Future developments in sensor nodes must produce very powerful and cost-effective devices, so that they may be used in applications like underwater acoustic sensor systems, sensing based cyber physical systems, time-critical applications, cognitive sensing and spectrum management, and security and privacy management. With the advances in the technology of microelectromechanical system (MEMS), developments in wireless communications and WSNs have also emerged. WSNs have become the one of the most interesting areas of research in the past few years. Here, we look into the recent advances and future trends in WSNs.

9.2 Applications and Opportunities

Wireless mobile ad hoc networks and sensor networks have been the focus of many recent research and development efforts. As the world of computing is getting portable and compact, the significance of MANET cannot be ignored. MANETs have the ability to set up networks on the fly in a harsh environments where it may not possible to deploy a traditional network infrastructure.

A decentralized network configuration is an operative advantage rather a necessity for military applications, and hence ad hoc pocket radio networks can be considered to deploy in army applications [26, 27]. Sensor networks have been proposed for a variety of applications, like intrusion detection and tracking for military purpose, habitat monitoring and motion diction for understanding earthquake patterns and prevention of theft, health applications by monitoring the drugs administered to patients, and for traffic analysis.

Wireless sensor networks use battery-operated computing and sensing devices [28]. A network of these devices will collaborate for common applications such as environmental monitoring, smart homes/offices, and avionics applications. We expect sensor networks to be deployed in an ad hoc fashion, with individual nodes remaining largely inactive for long duration of time but then becoming suddenly active when something is detected. These characteristics of sensor networks and applications motivate a MAC such as IEEE 802.11 in almost every way. So far, very few commercial sectors have come forward with equipment for wireless mobile computing and have not been available at a price attractive to large markets. But, with the increase of the capacity of mobile computers, the requirement for unlimited networking is also supposed to rise. Under some occasions due to natural disaster, no infrastructure (cellular or fixed) is available, and commercial ad hoc networks can be utilized. Citation of such commercial ad hoc networks includes rescue operation in some remote areas or when local coverage must be deployed as and when needed to serve as wireless public access in urban areas, providing faster deployment and extended coverage. Ad hoc networks linking notebooks or palmtop computers can be used to spread and share information among participants at a conference at the local level. They can also be deployed in home networks where devices are capable of communicating directly to exchange information like alarms, audio-video devices, and configuration updates. In this context, we can also consider that they will be appropriate for applications in autonomous networks of interconnected home robots that perform security surveillance, clean, do dishes, mow the lawn, and so on. Some researchers are even thinking of using them for environmental monitoring. Examples include forecasting water pollution and providing early warning of an approaching tsunami. There are wide ranges of applications of sensor networks determined mainly by application requirements, means of power supply, modes of deployment, or sensing modality.

9.3 Typical Applications

Wireless sensor networks and MANETs have been the focus of many recent development efforts because of numerous factors associated with business, regulation, technology, and social behavior.

Wireless networking, which includes wireless ad hoc and sensor networks, emerges from the integration of personal computing, cellular technology, and the Internet, because of the increasing interaction between computing and communication. Basically, they are changing information access from "anytime anywhere" into "all the time, everywhere." Such networks can be deployed whenever there is a necessity for establishing a networking environment for a limited duration of time. Wireless ad hoc network finds applications in the areas like home networks, vehicle networks, community networks, emergency response networks, and enterprise networks. These networks support innumerable occasions and can be deployed in a lot of situations, especially where infrastructure for communication is either difficult to set up within timing constraints or nonexistent. Some of such applications are given:

- Academic environment;
- Industrial or corporate environment;
- Healthcare;
- Defense (air force, navy, and army);
- Disaster situations for search and rescue.

Many more applications are there where we can use these networks.

9.3.1 Academic Environment

Currently, almost all academic institutions already have wireless communication networks to support a convenient environment among the students, research scholars, and faculty members for better interaction and better accomplishment of their mission. As an example, the lab instructor and the students enrolled under him for the lab can set up an ad hoc wireless communication network to provide an easy and convenient mechanism for the instructor to distribute handouts to students and for students to submit their assignments. Information sharing among the classmates can be as simple as a mouse click. Such networks can also be utilized when a group of students are on field trips or industrial visits.

9.3.2 Industrial or Corporate Environment

Wireless transmission of data has long been used in industrial applications, but recently it has gained importance. Successful use of wireless sensors in systems such as supervisory control and data acquisition has proven that these devices could effectively address the needs of industrial applications. The critical process applications of WSNs in industry are monitoring temperature, flow level, and pressure parameters.

Most businesses and corporations now have wireless communication networks in some form, especially in their manufacturing environments. Usually, manufacturing units have a lot of electronic devices (for monitoring several parameters/activities) that are interconnected. Such connections established through electric wires lead to busting and crowding of space, which may impose safety hazards and can significantly affect reliability. Deployment of some form of wireless networks can cast out many of these concerns. We can establish the connectivity among the various devices in the form of wireless ad hoc networks, which provide several attractive aspects including mobility. Under such networking situations, the devices can be easily relocated and the networks reconfigured, constrained by the needs of the devices. With the rapidly increasing technological advances in wireless technology and its subsequently decreasing prices, numerous wireless applications are being developed in industry. WSN in manufacturing industries can monitor and optimize quality control. Employees can organize corporate meetings without gathering in the same room, while communication among various entities can be maintained.

9.3.3 Health Care

Multimedia (data, video, and audio) content/information exchange between a patient and a hospital/physician can establish a better and immediate health care in critical and emergency situations. Using wireless ad hoc networks, a patient/attendant who is being transported to a hospital by an ambulance can exchange information. In many occasions, a doctor is in a much better position to diagnose and make a treatment plan for the patient if she has video information rather than just data and audio information, which may be useful in assessing the reflexes and viewing the coordination capability of a patient. Moreover, visual information of the patient can judge the level of injuries of the patient and can be treated in a much better way. A doctor can prepare a treatment plan for a patient who is being transported to a hospital before the patient's arrival at the hospital, if she gets a real time ultrasound scenes (or other video information) of the patient's heart, kidneys, or gallbladder. Wireless networks can be used to transmit such multimedia information from the ambulance to a hospital or to other health care professionals who are presently scattered at different places but are coming toward the hospital for the treatment of the patient being transported [27, 29].

9.3.4 Defense (Army, Navy, and Air Force)

There are several defense activity centers where communication infrastructure is not available. New and emerging technologies, such as networks, support military operations by delivering critical information rapidly and dependably to the right individual or organization at the right time. The new technologies must be integrated quickly into a comprehensive architecture to meet the requirements of present time. Other important applications are detection of enemy units' movements on land or sea, sensing intruders on bases, chemical or biological threats, and provision of logistics in urban warfare, command, control, communications, computing, intelligence, surveillance, reconnaissance, and targeting systems [12]. Military sensor networks detect and gain as much information as possible about enemy movements, explosions, and other phenomena of interest. One of the key aspects of any successful defense operations is secure communication. Infrastructureless setup of ad hoc networks can be used in military deployment and emergency operations. Defense planes or ships flying in a war zone may establish an ad hoc wireless network for communicating with each other for sharing images and data among themselves. In defense, information gathering is an important aspect where wireless sensor networks can be used for information-gathering purposes. Sensors for such applications are disposable and are used for an application once. These sensors are deployed either by air or by some appropriate means of gathering intelligence. They will remain suspended in the air for some time because of their tiny size, and they will gather intelligence/information for which they have been programmed, process the information, share among other nearby sensors, reach a consensus, and transmit information to a central location. The central processing facility then analyzes the transmitted information and takes a decision accordingly.

Target detection, an essential requirement in defense operations, can be accomplished using wireless sensors networks. Due to the advancement of semiconductor technology, a tiny chip is capable of mustering higher and higher processing power and can be used for the formation of smart dress, which defense people can wear in combat situations. Smart dress includes wireless network systems that can gather information, process it, and take necessary action based on the gathered information.

9.3.5 Disaster Situation for Search and Rescue Operations

Wireless networks are found to be very useful in search and rescue operations when we encounter an unfortunate situation like flood, hurricane, earthquake, or similar disaster that disrupts the power and communication capabilities by destroying the required infrastructures. In such disastrous situations, wireless networks can be established without infrastructures that support communication among various relief organizations for coordinating their activities. The number of robots to be used in the wireless networks for search and rescue operations depends on the size of the affected area by a disaster. Based on such deployment, proper action can be taken to incorporate the appropriate measures or help needed. Sensor networks to detect and monitor environmental changes in plains, valleys, forests, oceans, and so on.

9.3.6 Traffic Management and Monitoring

Every big city around the world suffers from traffic congestion. A sincere effort is being made to ease the traffic congestion. Congestion can be alleviated by planning managing traffic.

Real-time automatic traffic data collection must be employed for efficient management of rush hour traffic. Research on this topic is considered as part of the intelligent transport system (ITS) research community. ITS can be used in the application of computers, communications, and sensor technology to surface transportation.

The vehicle tracking application is to locate a specific vehicle or moving object and monitor its movement [15]. This work also describes design of WSN for vehicular monitoring. As the power source (battery) is limited, it is important that a design of sensor node should be power efficient.

9.3.7 Other Applications

9.3.7.1 Structural Healthcare

Structures are inspected at regular time intervals and repaired or replaced based on the time of use, rather than on their working conditions. The sensors embedded into structures enable condition based maintenance of these assets. Wireless sensing will allow assets to be inspected when the sensors indicate that there may be a problem. This will reduce the cost of maintenance and preventing harmful failure. These applications include sensors mounted on heavy-duty bridges, within concrete and composite materials, and big buildings.

9.3.7.2 Agriculture

Agriculture can also be benefited by the deployment of WSN to get information regarding soil degradation and water scarcity. With help of WSNs, we can check the clean water consumed in irrigation and manage it [30].

9.3.7.3 Smart Home/Office

Smart home environments can provide custom behaviors for a given individual. A considerable amount of research has been devoted to this topic. The research on smart homes is now starting to make its way into the market. It takes a considerable amount of work and planning to create a smart home. There are many examples of products currently on the market that can perform individual functions that are considered to be part of a smart home [26].

9.4 Primary Issues for Wireless Networks

Wireless communication (ad hoc/sensor networks) plays a very important role in our daily lives. An ad hoc/sensor network consists of autonomous self-organizing mobile devices that communicate with each other by creating a network in a given area. Each of these nodes is battery operated, and energy is exhaustible at very high rate. It is important to recognize the appropriate routing protocol that gets the work done with less energy consumption. Conserving energy is important in these scenarios so that the deployed network lasts for a long time. Battery life plays a very important role in establishing the network. Failure of battery life may lead to havoc in networks. Some of the nodes may die, leaving a hole in the network communication. Before establishing another route, the remaining nodes have to move within the communication range, thereby consuming energy that results in low battery. All these activities will deplete the energy available in the nodes very quickly. So, routing algorithms deployed in ad hoc/sensor networks play a key role in reducing the overhead involved in energy consumption, thus ensuring the longevity of the network. Hence, wireless sensor networks (WSNs) supported by recent technological advances in low-power wireless communications have been subject of panoptic study.

Improvements in hardware technology have resulted in low-cost sensor nodes, which are composed of a single chip embedded with memory, a processor, and a transceiver. These devices are multiprocessing sensors, which are able to process data and communicate in a wireless manner, within short distances, with each other. Low-power capabilities lead to limited coverage and communication range for sensor nodes compared to other mobile devices. The primary objective in wireless sensor network design is maximizing node/network lifetime, leaving the other performance matrices as secondary objectives. Since the communication of sensor nodes will be more energy consuming than their computation, it is a primary concern to minimize communication while achieving the desired network operation. Topology control has great importance for prolonging lifetimes, reducing radio interference, and increasing efficiency of media access control protocols and routing protocols, thereby ensuring quality of connectivity and coverage and increasing the network service.

QoS support is challenging due to severe energy and computational resource constrains of wireless sensors [31]. Various service properties such as the delay, reliability, network lifetime, and quality of data may conflict. For example, multipath routing can improve reliability; however, it can increase the energy consumption and delay due to duplicate transmissions. Modeling such relationships, measuring the provided quality, and providing means to control the balance is essential for QoS support.

9.5 Recent Developments

Advances in computing and communication technologies have been stimulated due to the availability of faster, cheaper, and more reliable electronic components. Several theoretical and algorithmic schemes (more will also be developed in near future) that incorporate key issues in SAP networks (which include sensor networks, ad hoc wireless networks, and peer-to-peer networks) carry out a central role in the development of emergency network paradigms. The central technical issues of SAP networks are to be explored and probable solution or tools are to be searched to address these issues. One of the major and most significant challenges in wireless communication network is how to improve the overall network throughput while maintaining low energy consumption for packet processing and communication.

In recent years there have been growing interests in mobile communications and wireless network technologies. Recent advancements in the field of sensing, computing, and communications have attracted research efforts and huge investments from various quarters in the field of wireless ad hoc and sensor networks. The various areas where major research activities going on in the field of WSN are deployment, localization, synchronization, data aggregation, dissemination, database querying, architecture, middleware, security, designing low-power-consumption devices, abstractions, and high level algorithms for sensor specific issues.

Some research issues that can be considered are different strategies to improve signal reception; design of low-power, less cost sensors; and processing units [32]. Various plans to conserve node power consumption and node optimization, and simple modulation schemes may also be considered for sensors.

9.6 Active Research Areas

A wireless communication network that includes ad hoc and sensor networks represents technological miracles. There are several challenges that need to be taken care for fully exploring their benefits. Wireless communication networks have several constraints (which are also to be addressed) like communication overhead, limited bandwidth and capacity, limited battery power and life, size of the mobile devices, and data security.

Wireless sensor networks use battery-operated computing and sensing devices. A network of these devices will collaborate for a common application, such as environmental monitoring. We expect sensor networks to be deployed in an ad hoc fashion, with individual nodes remaining largely inactive for a long duration of time, but then becoming suddenly active when something is detected. These characteristics of sensor networks and applications motivate a MAC that is different from traditional wireless MAC, such as IEEE 802.11 in almost every way: energy conservation and self-configuration are primary goals, while per-node fairness and latency are less important. S-MAC uses three novel techniques to reduce energy consumption and support self-configuration. Although capacity can be enhanced by deploying efficient transmission techniques, it is still inadequate and hence requires the implementation of some innovative/renovative approaches for optimal utilization of available bandwidth and capacity. Goal-oriented intensive research is still needed to give more efficient mechanisms for using the available communication bandwidth in a wireless communication environment. Nowadays more and more electronic components can be placed on smaller chips because of the advancements in semiconductor technologies, thereby developing mobile devices that are more powerful and less power hungry. Due to the shrinkage of mobile devices, it has become possible to incorporate more and more features and functionalities in those devices without much demand of power. The challenge now is to maintain the trend. Wireless networks are more prone to security risks compared to other networks.

The field that paid less attention is the privacy concerns on information being collected, transmitted, and analyzed in a WSN [33]. Such private information of concern may include payload data collected by sensors and transmitted through the network to a centralized data processing server. The location of a sensor initiating data communication, and other such context information, may also be the focus of privacy concerns.

In real-world applications of WSNs, effective countermeasures against the disclosure of both data and context oriented private information are indispensable prerequisites. Privacy protection is needed in various fields related to WSNs, such as wired and wireless networking, databases, and data mining. Effective privacy preserving techniques are needed for the unique challenges of WSN security. But the addition of any level of information security will incorporate additional overhead, thereby requiring additional bandwidth for transmission. Extensive research is going on to discover techniques to support secure information transfer and at the same time will not add prohibitive overhead.

Recent research advances in low-power, low-cost, and high-data-rate wireless communications endure a promising future for the deployment of sensor networks to support a broad range of applications like health monitoring, habitat monitoring, target tracking, and disaster management. The performance of sensor networks is closely related to the rate at which the source nodes generate the traffic. Care should be taken that the generated traffic does not congest and degrade the performance of the network. Thus, it is necessary to model the network's traffic characteristics accurately in order to develop efficient protocols for wireless sensor networks. WSNs have met a huge growth and significant future prospects of evolution, meeting applications from medical, environmental surveillance, robotics, military, smart vehicles, and domestic areas. The main reason for this growth are the high fault tolerance, fast deployment, and self-organizing capabilities of WSNs, as well as their low cost and high density of deployment, which does not affect the functionality of the application when sensor nodes fail or are destroyed. Other business applications of WSNs include climate control in buildings and interactive games. Hence, for example, in target tracking and boarder surveillance applications, sensor networks must include a large number of nodes in order to cover the target area successfully.

A significant progress in research can be seen in WSNs topology control. Many topology control algorithms have been developed to date, but problems such as lack of definite and practical algorithms, lack of efficient measurement of network performance, and practicality of mathematical models still exist. Several graph models used in topology control, the present hot spots, and the future trends on the research of topology control.

In the future, variation of energy consumption by mobility models may be incorporated in the model and can be studied on the variation of mode density, traffic load, number of sources, transmission range, and network size. An investigation can also be made to study the effect of mobility, traffic, source, and network on the energy overhead of various routing protocols. Future scope includes designing a mathematical model for maximum allowable sink mobility. With all these research scope and challenges, we firmly believe that we have a very exciting time ahead of us in the area of wireless sensor networks.

9.7 Challenges and Future Scope

Research areas include designing low-power-consumption communication and complementary metal oxide semiconductor circuit techniques specifically optimized for sensor networks, designing new architecture for integrated wireless sensor systems, and modulation method and data rate selection [34]. The research on mobile nodes localization and motion analysis in real time will continue to grow as sensor networks are deployed in large numbers and as application becomes varied. Scientists in numerous disciplines are interested in methods for tracking the movements and population control of animals in their habitats. Another important application is to design a system to track the location of valuable assets in an indoor environment. We need to improve the maximum likelihood estimation in a distributed environment like sensor networks. Developing mobile-assisted localization is another important research area. One needs to improve the localization accuracy.

Sensor networks are still at an early stage in terms of technology, as it is still not widely deployed in the real world and this opens many doors for research. The current routing protocols needs to be improved, as they have their own set of problems. Much work is not reported on contention issues or high network traffic. Developing transport protocols for sensor networks is itself a difficult task, and not much work is reported. Existing transport layer protocols for WSNs assume that the network uses a single path routing, and multipath routing is not considered, which opens many doors for research in this direction. Many transport protocols do not consider priority when routing. Since sensor nodes are placed in various types of environments, data from different locations will have different priorities.

The main research focus in data aggregation is focused toward conserving energy. Other research issues include improving security in data aggregation (i.e., tradeoffs between different objectives such as energy consumption, latency and data accuracy, improving quality of service of the data aggregation protocols in terms of the band width, and end-to-end delay).

Some research areas in sensor database include providing spatio-temporal querying, multiquery optimization, storage placements, designing distributed long-term networked data storage, enabling low-energy communication overhead, various ways of representing the sensor data, processing and distributing query fragments, dealing with communication failures, and designing various models for deploying and managing a sensor database systems [35]. The design and implementation of a middleware layer for fully realizing the potential of wireless sensor network is an open research area that still needs to be investigated further. One needs to design developer-friendly middleware architecture that is not only generic but also should take care of all the underlying hardware intricacies while helping to reduce energy consumption and provide adequate quality of support. The area of sensor network's QoS remains largely an unexplored research area: designing an appropriate QoS model, deciding how many layers need to be integrated, providing support for heterogeneous nodes, designing QoS model for specific applications, designing QoS via middleware layer, and designing QoS based protocols to integrate them with other networks like cellular, LANs, and IP.

The security issues posed by sensor networks are a rich, interesting field for research problems, such as designing routing protocols having built-in security features, a new symmetric key cryptography for sensor networks, secure data aggregation protocols, intrusion detection systems, and security systems for multimedia sensors.

9.7.1 Research Challenges

9.7.7.1 Power

Power is always been a challenge for WSNs designs. One of the ways to prolong the network lifetime is to design energy-efficient algorithms and hardware that use power intelligently [36–38].

9.7.7.2 Hardware Cost

One of the main challenges is to produce low-cost and tiny sensor nodes. Current sensor nodes are mainly prototypes with respect to these objectives. Low cost of sensor nodes can be achieved by recent and future progress in the fields of MEMS.

9.7.7.3 Security

Security is one of the major challenges in WSNs. Most of the attacks that are performed on WSN are insertion of false information by compromised nodes within the networks. Development of security schemes for WSN also faces challenges related to constrained environment.

9.7.7.4 System Architecture

Research in the field of WSN is going on around the world, but still there is no unified system and network architecture on the top of which different application can be built.

9.7.7.5 Real-World Protocols

Protocols need to be developed for real-world problems considering the theoretical concepts and synthesizing novel solutions into a complete systemwide protocols for real-world application.

9.7.7.6 Analytical and Practical Results

To date, very few analytical results exists for WSNs. All new applications get confidence only when tested and analyzed practically and results compared with existing schemes.

The inherent nature of wireless networks makes them deployable in a variety of circumstances. They have the potential to be everywhere, on roads, in our homes and offices, forests, battlefields, disaster-struck areas, and even under water in oceans. This paper surveys the application areas where wireless networks have been deployed, such as military sensing, traffic surveillance, target tracking, environment monitoring, and health care monitoring as summarized in Table 9.1 and Table 9.2 [39]. The paper also surveys the various fields where

Table 9.1
Recent Advances and Future Trends of MANETs

Area	Applications
Context-aware services	Information services: advertise location specific services (e.g., gas stations), time dependent services (e.g., printer, phone, server, fax, gas stations) availability information. Infotainment: touristic information. Follow-on services: automatic call-forwarding, mobile workspace.
Tactical networks	Automated warfield Military communication and operations
Home and enterprise networking	Home/office wireless networking (e.g., shared whiteboard application) Conferences and campus settings Set up ad hoc communication during conferences, meetings, and lectures
Commercial and civilian environments	E-commerce: electronic payments from anytime and anywhere Sports stadiums, trade fairs, shopping malls Business: dynamic access to customer files stored in a central location on the fly, provide consistent databases for all agents, mobile office Vehicular services: road or accident guidance, transmission of road and weather conditions, intervehicle network, taxi cab network
Emergency services	Search and rescue operations, as well as disaster recovery (e.g., early retrieval and transmission of patient data [records, status, diagnosis] from/to the hospital) Replacement of a fixed infrastructure in case of earthquake, hurricanes, fire, and so on Supporting doctors and nurses in hospitals
Educational applications	Set up virtual class rooms or conference rooms Universities and campus settings Set up ad hoc communication during conferences, meetings, and lectures
Entertainment	Multiuser games Wireless P2P networking Robotic pets Theme parks Outdoor Internet access
Coverage extension	Extending cellular network access Linking up with the Internet, intranets and so on
Sensor networks	Items listed in Table 9.2 (i.e., meant for WSN)

Table 9.2
Recent Advances and Future Trends of WSNs

Area	Applications
Smart home/office	Provides custom behavior
Military	Military situation awareness, battlefield surveillance, sensing intruders, detection of enemy units
Industrial and commercial	Monitoring and control of industrial equipment, manufacturing monitoring
Traffic management and monitoring	Traffic congestion control
Structural healthcare	Condition based maintenance, concrete and composite materials
Topology and coverage control	Prolong lifetime, reducing radio interference, increasing the efficiency, graph models of topology control
Quality of service provision	Tradeoff between reliability and energy consumption, delay energy consumption, and delay
Mobility management	NEMO, MANEMO
Security and privacy concern	Security in wired and wireless networking, databases, and data mining
Biomedical/medical	Code Blue, ALARMNET, AMON, Gluco Watch G2
Cognitive sensing	Bio-inspired sensing, swarm intelligence, Quorum sensing
Spectrum management	Multiple frequencies for parallel communication, self-adaptive spectrum management
Underwater acoustic sensor systems	Oceanographic data collection, pollution monitoring, disaster prevention, assisted navigation, tactical surveillance
Coordination in heterogeneous	Connecting the WSN with heterogeneous network, gateway based interconnection, and overlay based interconnection
Sensing based cyberphysical systems	Collision avoidance, robotic surgery, and nano level manufacturing, search and rescue operation, air traffic control, healthcare monitoring
Time-critical applications	Real-time applications like fire monitoring, border surveillance, medical care, and highway traffic coordination
Experimental systems and new applications	WSNs for oil & gas, monitoring wildlife passages, fire hazard monitoring, NEURON, wireless multimedia sensor networks
New models and architectures	EAWNA, cubic and cross layer (CCL), WASP

wireless networks may be deployed in the near future as underwater acoustic sensor systems, sensing based cyber physical systems, time-critical applications, cognitive sensing and spectrum management, and security and privacy management. These application areas are being researched extensively by various people across the industry and academia.

References

[1] Akyildiz, I. F., W. Su, Y. Sankarasubramaniam, and E. Cayirci, "A Survey on Sensor Networks," *IEEE Communications*, 2002, pp. 102–114.

[2] Tiwari, A., A. Lewis, and S. G. Shuzhi, "Design and Implementation of Wireless Sensor Network for Machine Condition Based Maintenance," Int'l Conf. Control, Automation, Robotics, & Vision (ICARV), Kunming, China, December 6–9, 2004.

[3] Arms, S. W., C. P. Townsend, and M. J. Hamel, "Validation of Remotely Powered and Interrogated Sensing Networks for Composite Cure Monitoring," 8th International Conference on Composites Engineering (ICCE/8), August 7–11, 2001.

[4] Jardosh, S., and P. Ranjan, "A Survey: Topology Control for Wireless Sensor Networks," International Conference on, Signal Processing, Communications and Networking, 2008, pp. 422–427.

[5] Dinh, T. N., Y. Xuan, M. T. Thai, P. Pardalos, and T. Znati, "On New Approaches of Assessing Network Vulnerability: Hardness and Approximation," *IEEE/ACM Transactions on Networking (ToN)*, 2011, in press.

[6] Xuan, Y., I. Shin, M. T. Thai, and T. Znati, "Detecting Application Denial-of-Service Attacks: A Group Testing Based Approach," *IEEE Transactions on Parallel and Distributed Systems*, Vol. 21, No. 8, 2010, pp. 1203–1216.

[7] Thai, M. T., R. Tiwari, and D.-Z. Du, "On Construction of Virtual Backbone in Wireless Ad Hoc Networks with Unidirectional Links," *IEEE Transactions on Mobile Computing*, Vol. 7, No. 8, 2008, pp. 1–12.

[8] Pease, S. G., L. Guan, I. Phillips, A. Grigg, "Cross Layer Signalling and Middleware: A Survey for Inelastic Soft Real-time Applications in MANETs," *Elsevier Journal of Network and Computer Applications*, 2011, in press.

[9] Camilo, T., P. Pinto, A. Rodrigues, J. SaSilva and F. Boavida, "Mobility Management in IP Based Wireless Sensor Networks," *World of Wireless, Mobile and Multimedia Networks*, Vol. 23, No. 26, 2008, pp. 18.

[10] Liutkevicius, A., A. Vrubliauskas, and E. Kazanavicius, "A Survey of Wireless Sensor Network Interconnection to External Networks," *Novel Algorithms and Techniques in Telecommunications and Networking*, 2010, pp. 41–46.

[11] Dai, H., and E. Chan, "Quick Patching: An Overlay Multicast Scheme for Supporting Video on Demand in Wireless Networks," *Multimedia Tools and Applications*, Vol. 36, No. 3, 2008, pp. 221–242.

[12] Younis, O., et al., "Cognitive Tactical Network Models," *IEEE Communications*, October 2010.

[13] Younis, O., L. Kant, K. Chang, K. Young, and C. Graff, "Cognitive MANET Design for Mission-Critical Networks," *IEEE Communications*, Vol. 47, No. 10, October 2009, pp. 64–71.

[14] Biradar, R. C., and S. S. Manvi, "Link Stability Based Multicast Routing in MANETs," *Elsevier International Journal on Computer Networks*, Vol. 54, No. 7, May 2010, pp. 1183–1196.

[15] Chinrungrueng, J., U. Sununtachaikul, and S. Triamlumlerd, "A Vehicular Monitoring System with Power Efficient Wireless Sensor Networks." In *Proc. 6th International Conference on ITS Telecommunications*, 2006, pp. 951–954.

[16] Wang, N., M. H. Wang, and N. Q. Zhang, "Wireless Sensors in Agriculture and Food Industry: Recent Development and Future Perspective," *Computers and Electronics in Agriculture*, Vol. 50, No. 1, 2006, pp. 1–14.

[17] Akyildiz, I. F., D. Pompili, and T. Melodia, "Underwater Acoustic Sensor Networks: Research Challenges," *Ad Hoc Networks*, Vol. 3, 2005, pp. 257–279.

[18] Garcia-Sanchez, A. J., F. Garcia-Sanchez, F. Losilla, P. Kulakowski, and J. G. Haro, et al., "Wireless Sensor Network Deployment for Monitoring Wildlife Passages Sensors," *Sensors 2010*, Vol. 10, No. 8, 2010, pp. 7236–7262.

[19] Zeng, Y., N. Xiong, J. Hyuk Park, and G. Zheng, "An Emergency Adaptive Routing Scheme for Wireless Sensor Networks for Building Fire Hazard Monitoring Sensors," *Sensors 2010*, Vol.10, No. 6, 2010, pp. 6128–6148.

[20] Li, W., E. Chan, S. Lu, and D. Chen, "Communication Cost Minimization in Wireless Sensor and Actor Networks for Road Surveillance," *IEEE Trans. on Vehicular Technology*, in press.

[21] Patel, R. B., D. Kumar, and T. C. Aseri, "Multi-Hop Communication Routing (MCR) Protocol for Heterogeneous Wireless Sensor Networks," *International Journal Information Technology, Communication and Convergence Computer Systems Science and Engineering*, Vol. 1, No. 2, 2010, pp. 1–14.

[22] Patel, R. B., K. Kumar, and A. K. Verma, "A Location Dependent Connectivity Guarantee Key Management Scheme for Heterogeneous Wireless Sensor Networks," *Journal of Advances in Information Technology*, Vol. 1, No. 3, August 2010, pp. 105–115.

[23] Patel, R. B., D. Prasad, and Manik Gupta, "SEEAR-II: A System Model for Secure and Energy Efficient Adaptive Routing in Wireless Sensor Networks," *International Journal of Computer Science & Management Systems*, Vol. 3, No. 1, June 2011, pp. 27–39.

[24] Patel, R. B., K. K. Sharma, and H. Singh, "A Hop by Hop Congestion Control Protocol to Mitigate Traffic Contention in Wireless Sensor Networks," *International Journal of Computer Theory and Engineering*, Vol. 2, No. 6, December 2010, pp. 1793–8201.

[25] Wang, C., M. T. Thai, Y. Li, F. Wang, and W. Wu, "Optimization Scheme for Sensor Coverage Scheduling with Bandwidth Constraints," *Optimization Letters*, Vol. 3, No. 1, 2009, pp. 63–75.

[26] Wood, A., G. Virone, T. Doan, Q. Cao, L. Selvao, et al., "ALARMNET: Wireless Sensor Networks for Assisted Living and Residential Monitoring" *SiteSeerX*, 2006, pp. 1–14.

[27] Anliker, U., J. A. Ward, P. Lukowicz, G. Tröster, and F. Dolveck, et al., "AMON: A Wearable Multi Parameter Medical Monitoring and Alert System," *IEEE Transactions on Information TechNology in Biomedicine*, Vol. 8, No. 4, 2004, pp. 415–427.

[28] Guan, X., L. Guan, and X. Wang, "A Novel Energy Efficient Clustering Technique Based on Virtual Hexagon For Wireless Sensor Networks," *International Journal of Innovative Computing, Information, and Control*, Vol. 7, No. 2, 2011.

[29] Iafusco, D., M. K. Errico, C. Gemma, et al., "Usefulness or Uselessness of GlucoWatch in Monitoring Hypoglycemia in Children and Adolescents," *Pediatrics*, Vol. 113, No. 1, 2004, pp. 175–176.

[30] Shen, C. C., C. Srisathapornphat, and C. Jaikaeo, "Sensor Information Networking Architecture and Applications" *IEEE Personal Communications*, 2001, pp. 52–59.

[31] Abidin, H. Z., and F. Y. A. Rahman, "Provisioning QoS in Wireless Sensor Networks Using a Simple MaxMin Fair Bandwidth Allocation," *World Congress on Computer Science and Information Engineering*, Vol.1, 2009, pp. 44–48.

[32] Doumit, S. S., and D. P. Agrawal, "Self Organizing and Energy Efficient Network of Sensors" *IEEE*, 2002, pp. 1–6.

[33] Li, N., N. Zhang, S. K. Das, and B. Thuraisingham, "Privacy Preservation in Wireless Sensor Networks: A State of the Art Survey," *Ad Hoc Networks*, Vol. 7, 2009, pp. 1501–1514.

[34] Patel, R.B., and D. Kumar, "A Novel Multihop Energy Efficient Heterogeneous Clustered Scheme for Wireless Sensor Networks," *Tamkang Journal of Science and Engineering*, Vol. 14, 2011.

[35] Liu, L., "Research on Environment Adaptive Architecture Model of Wireless Sensor Networks." In *Proc. 2nd International Conference on Networks Security, Wireless Communications and Trusted Computing*, 2010, pp 130–133.

[36] Li, H., E. Chan, and G. Chen, "AEETC: Adaptive Energy-Efficient Timing Control in Wireless Networks with Network Coding," *Telecommunication System*s, Vol. 45, No. 4, December 2010, pp. 289–301.

[37] Chan, E., and S. Han, "Energy Efficient Residual Energy Monitoring in Wireless Sensor Networks," *International Journal of Distributed Sensor Networks* (in press).

[38] Ye, M., E. Chan, G. Chen, and J. Wu, "Energy Efficient Fractional Coverage Schemes for Low Cost Wireless Sensor Networks," *International Journal of Sensor Networks* (in press).

[39] Hoebeke, J., I. Moerman, B. Dhoedt, and P. Demeester, "An Overview of Mobile Ad Hoc Networks: Applications and Challenges," Technical Paper, Department of Information Technology (INTEC) Ghent University-IMEC VZW, Sint Pietersienuivstraat 41, B-9000 Ghent, Belgium, No. 4, p. 60.

About the Author

Professor Subir Kumar Sarkar received his B.Tech, M.Tech, and Ph.D. degrees from the Institute of Radio Physics and Electronics, University of Calcutta, India, and post-doctoral fellow from the Deptartment of Electrical and Computer Engineering at Virginia Commonwealth University.

He served as an executive engineer at Oil and Natural Gas Corporation Ltd. (ONGC) for about 10 years before joining the teaching profession. He joined the faculty of the Deptartment of Electronics and Telecommunication Engineering at Bengal Engineering and Science University, Shibpur, India in April 1992. In 1999 he joined the same department at Jadavpur University, Kolkata, India, and currently he is a professor. He has published three engineering textbooks, and few are in the pipeline.

He has published more than 360 technical research papers in archival international and national journals and peer-reviewed conferences in the field of devices, mobile ad hoc and sensor networks, digital watermarking, and RFID and its applications.

To date, 26 scholars have been awarded Ph.D. (Eng.) degrees under his guidance, and eight more are currently pursuing their research leading to a Ph.D.

He is technical program committee member for several major international and national conferences and symposiums and was the organizing secretary of the 2004 international conference on Communications, Devices, and Intelligent Systems (CODIS2004).

As a principal investigator, he has successfully completed six research and development projects and four more are currently ongoing.

He has visited several countries, including France, the United Kingdom, Switzerland, and Japan, for training, presenting papers, and visiting sophisticated laboratories. His most recent research focuses are in the areas of ad hoc networks, wireless sensor networks, mobile communications, RFID and its applications, and digital watermarking and data security.

He is a senior member of IEEE, life fellow of the Institution of Engineers, life fellow of Institution of Electronics and Telecommunication Engineers, life member of ISTE, and life member of the Indian Association for the Cultivation of Science.

Index

Academic environment applications, 234
Adaptive core multicast routing (ACMR) protocol, 25
Adaptive routing protocols, 150
Ad hoc demand driven multicast routing (ADMR), 25
Ad hoc Internet access routers (AIARs), 123
Ad hoc on-demand distance vector (AODV), 24, 47–48
    defined, 47
    OLSR comparison, 25
    performance analysis, 43
    RERR packets, 47–48
    route establishment, 48
    RREQ packets, 47–48
    secured (SAODV), 26
Ad hoc on-demand distance vector routing algorithm by Uppsala University (AODVUU), 48–49
Ad hoc on-demand multipath distance vector (AOMDV), 42
    average energy consumed versus mobility speed, 71, 72, 73
    average energy consumed versus traffic load, 82, 83
    average network delay versus traffic load, 78
    average network delay versus with mobility speed, 65, 66, 67
    connectivity, 126
    defined, 49
    multiple loop-free and link disjoint path technique, 49
    network throughput versus mobility speed, 67, 68, 69
    network throughput versus traffic load, 79, 80
    PDR versus mobility speed, 64, 65
    PDR versus traffic load for, 77
    routing overhead versus mobility speed, 69, 70, 71
    routing overhead versus traffic load for, 80, 81
Ad hoc robust header compression (ARHC), 121
Applications
    academic environment, 234
    agriculture, 237
    Bluetooth, 13–14
    defense, 235–36
    health care, 234–35
    industrial/corporate environment, 234
    introduction to, 231–32
    MANETs, 26–28
    opportunities and, 232–33
    search and rescue operations, 236
    smart home/office, 237
    structural healthcare, 236–37
    traffic management/monitoring, 236
    typical, 233–37
    WiMAX, 10–11
    WSNs, 135–36, 233–37

Authentication, 164
Average end-to-end delay (AEED)
    number of actor nodes versus, 170
    performance evaluation, 169–70
    simulation time versus, 170
Average energy consumed
    defined, 62
    evaluation of, 71–73, 81–83
    Manhattan mobility model, 103
    with mobility speed, 71–73
    number of nodes versus, 75
    patrimonial ZigBee routing protocols, 224–25
    RWP-SS, 103
    with traffic load, 81–83
Average hop count, 103
Average network delay
    defined, 61
    evaluation of, 65–66, 76–78
    IEEE 802.15.4 network scenarios, 204–5
    with mobility speed, 65–66
    number of nodes versus, 74, 205
    number of sources versus, 205
    patrimonial ZigBee routing protocols, 220–21, 227
    RWP-SS, 100–101
    simulation time versus, 201
    with traffic load, 76–78
    traffic load versus, 204

Bansal energy model, 112, 123
    energy exhausted versus transmission range, 126
    values, 124
Berkeley motes, 137
Bluetooth, 3, 13–14
    applications, 13–14
    illustrated, 12
    for low-power consumption, 14
    uses, 13
Broadcasting, 31–33
    classifications, 31
    cluster-based, 32
    defined, 31
    optimized, 31–32
    storm problem, 32

Calibration, 147–48
CBR on-off traffic, 215

CDMA2000, 8
Cellular networks, 7–11
    advantage of, 7
    defined, 7
    4G, 7–9
    illustrated, 8
    WiMAX, 9–11
Chandrakasan energy model, 112, 124
    energy exhausted versus transmission range, 127
    values, 124
Chi Squared group force model, 95
City section mobility model, 43
Clock synchronization, 146–47
Cluster head nodes, 115
Clustering
    maintenance overhead, 119
    overhead minimization and analysis using, 116–20
Cluster message overhead, 116
Code division multiple access (CDMA), 5
Community-based mobility model
    average hop count, 103
    betweeness of edges, 97
    intercommunity edges, 97
    network throughput, 101
    PDR, 100
    routing overhead, 102
    social attractivity, 98
    social network input, 97
    steps in establishment, 98
Confidentiality, 164
Control messages, 116

Data aggregation
    accomplishment, 154
    cluster-based, 154–55
    design issues, 155
    protocols, 154
Data-centric storage (DCS), 158
Data dissemination, 156
Data flow, 156–57
Data transmission, IEEE 802.15.4, 188–89
Defense applications, 235–36
Delay energy-aware routing protocol (DEAP), 135
Deployment, 144–45
Destination-sequenced distance-vector (DSDV), 25, 51–52
    analysis, 45

defined, 51
full dump, 52
incremental updates, 52
node advertisement, 51–52
Directed diffusion, 151
Direction of arrival (DoA), 146
Direct sequence spread spectrum (DSSS), 140
Distributed Bellman Ford (DBF), 45
Dynamic MANET on-demand (DYMO), 42, 52–53
    average energy consumed versus mobility speed, 72
    average energy consumed versus traffic load, 82
    average network delay versus traffic load, 77
    average network delay versus with mobility speed, 66
    defined, 52
    energy consumption, 130
    network throughput versus mobility speed, 68
    network throughput versus traffic load, 79
    PDR versus mobility speed, 64
    PDR versus traffic load for, 76
    RERR messages, 53
    route breakage scenario, 53
    routing overhead versus mobility speed, 70
    routing overhead versus traffic load, 80
    RREQ messages, 52–53
Dynamic source routing (DSR), 25, 50–51
    defined, 50
    packet confirmations, 51
    RERR messages, 51
    routing establishment, 50
    RREQ messages, 50

Energy models, 123–24
    Bansal, 112, 123
    Chandrakasan, 124
    introduction to, 113
    literature background, 113–14
    performance evaluation of, 125–27
    simulation, 124–25
    Vaddina, 112, 123
Expected zone, 56
Exponential on-off traffic, 214–15, 216

Fisher-Snedecor group force model, 95
Fisheye state routing (FSR), 24, 25, 42, 54–55
    average energy consumed versus mobility speed, 72
    average energy consumed versus traffic load, 82
    average network delay versus traffic load, 77
    average network delay versus with mobility speed, 66
    defined, 54
    demonstration, 54
    network throughput versus mobility speed, 68
    network throughput versus traffic load, 79
    PDR versus mobility speed, 64
    PDR versus traffic load, 76
    routing overhead versus mobility speed, 70
    routing overhead versus traffic load, 80
    scope, 54
Flooding, 156
4G cellular networks, 7–9
Frequency hopping spread spectrum (FHSS), 140

Gateway advertisements (GWADV), 25
Gateway discovery, 122
Gauss Markov mobility model, 43
    defined, 93
    to mimic other models, 94
    node velocity, 93–94
General packet radio service (GPRS), 5
Geographic hash tables (GHT), 158
Georgia Tech Network Simulator (GTNetS), 44
Global system for mobile communication (GSM), 5
Greedy-face-greedy (GFG) routing protocol, 158
Group force mobility model (GFMM), 87
    Chi Squared, 95
    defined, 94
    Fisher-Snedecor, 95
    loose group, 94
    PDR, 102
    Rayleigh, 95
    routing overhead, 102

Group force mobility model (continued)
　steps in establishment, 95
　tight group, 94

Header compression, overhead minimization by, 119–20
Health care applications, 234–35
Hello overhead, 116
Hierarchical routing scheme
　communication overhead, 115
　control overhead, 115
　overhead analysis, 114–15
　routing overhead reduction, 115
Hop-by-hop context initialization algorithm, 121
Hop-count based routing (HBR), 123

IEEE 802.11, 3
　defined, 5–6
　medium access control (MAC) layer, 28–29
　modulation techniques, 6
　specifications, 6
　standards, 6–7
IEEE 802.15.4
　association/disassociation, 193–94
　characteristics of, 184–94
　CSMA-CA mechanism, 183–84
　data gathering paradigm, 194–96
　data transmission, 188–89
　defined, 184
　IEEE 802.11 performance comparison, 182
　introduction to, 181–82
　literature background, 182–84
　LR-WPAN protocol architecture, 186–87
　MAC protocol, 182
　metrics evaluation with network scenarios, 201–8
　metrics evaluation with simulation time, 200–201
　metrics evaluation with traffic load, 198–200
　network topologies, 185–86
　on NS2 simulator, 182, 183
　overview of, 184–94
　PAN coordinator, 184, 185
　PAN identifier, 185
　PANs, starting/maintaining, 192–93
　performance analysis, 181–208
　products based on, 183
　result analysis, 197
　sensor network evaluation, 183
　simulation environment, 196–97
　simulation parameters, 197, 198
　superframe structure, 187–88
　synchronization, 194
IEEE 802.16 standard, 9
Industrial/corporate environment applications, 234
In-network, 157
Intercluster routing, 117
Internet connectivity
　MANETs, 34–35
　overhead minimization for ad hoc networks, 121–23

Levy walk mobility model, 96–97
　angle model, 97
　defined, 96
　flight, 96
　variables, 97
Localization, 145–46
Location-aided routing protocol (LAR), 42, 55–57
　average energy consumed versus mobility speed, 71
　average network delay versus with mobility speed, 65
　defined, 55
　expected zone, 56
　network throughput versus mobility speed, 67
　node location information, 55
　PDR, 63–64
　PDR versus mobility speed, 65
　request zone, 56
　route establishment, 55
　routing overhead versus mobility speed, 69
　RREQ messages, 56–57
　scheme illustrations, 57
Long term evolution (LTE), 9
Loose group, 94
Low energy adaptive cluster hierarchy (LEACH), 150–51
　defined, 150
　operation of, 151
　performance metrics, 213
LR-WPAN device architecture, 186–87

MAC protocols
    classification of, 142–43
    collision avoidance, 141
    contention-based, 142
    design issues of, 141–43
    EMAC, 142
    event reporting, 141–42
    message passing, 141
    node movement and, 143
    real-time requirements, 142
    S-MAC, 142, 143
    SMACS, 142
    T-MAC, 142
M-adaptive multidomain power-aware sensors (μAMPS) project, 138
Manhattan mobility model, 43
    average network delay, 103
    defined, 96
    energy consumed, 103
    parameters, 96
MANTIS operating system, 139
Medium access control (MAC) layer
    functionalities, 28
    IEEE 802.11, 28–29
Middleware
    design and implementation, 162
    issues, 161–62
    systems, 162
    WSNs, 161–62
Minimal benefit average, 122
Mobile ad hoc networks (MANETs)
    applicability of, 27
    applications of, 26–28
    attack types, 36
    authentication, 35
    availability, 35
    broadcasting, 31–33
    characteristics of, 27–28
    communication between nodes, 24
    confidentiality, 35
    defined, 11
    devices, 23
    fundamentals of, 23
    integrity, 35
    Internet connectivity for, 34–35
    literature background, 24–26
    MAC layer, 28–29
    mobility models, 87–107
    multicasting, 33
    nonrepudiation, 35
    overhead control mechanism, 111
    performance factors, 30–33
    routing protocols, 30–31, 41–83
    security in, 35–36
    topology control, 29–30
Mobile IP, 122, 123
Mobility, 88
Mobility models, 87–107
    city section, 43
    classification of, 91
    community-based, 97–98
    comparison, 43
    description of, 90–99
    Gauss Markov model, 43, 93–94
    group force (GFMM), 87, 94–95
    introduction to, 87–89
    Levy walk, 96–97
    literature background, 89
    Manhattan, 43, 96
    performance metrics for network scenarios, 103–7
    random direction, 43
    random walk with reflection (RW-R), 87, 92
    random walk with wrapping (RW-W), 87, 93
    random waypoint (RWP), 43, 87, 90
    random waypoint-steady state (RWP-SS), 87, 88, 90–92
    reference point group mobility (RPGM), 87, 94
    in routing protocol simulation, 41
    semi-Markov smooth (SMS), 98–99
    simulation environment, 99–100
    virtual track, 44
Mobility speed, 62
    average energy consumed with, 71–73
    average network delay with, 65–66
    network throughput with, 66–69
    PDR with, 63–65
    routing overhead with, 69–71
Motes, 136, 137
Multiactor/multisensor (MAMS) model, 135
Multicasting, 33
Multihop routers, 123
Multiple disjoint trees multicast routing protocol (MDTMR), 25
Multiple loop-free and link disjoint path technique, 49
Multipoint relay (MPR), 58

Nano-Qplus, 139
Negotiation-based routing protocols, 150
Network simulator, 60–61
Network throughput
  community model, 101
  defined, 61–62
  evaluation of, 66–69, 78–79
  IEEE 802.15.4 network scenarios, 205–6
  with mobility speed, 66–69
  number of actor nodes versus, 172
  number of nodes versus, 74
  number of sensor nodes versus, 206
  number of sources versus, 206
  patrimonial ZigBee routing protocols, 221–22, 228
  simulation time versus, 173, 202
  with traffic load, 78–79
  traffic load versus, 206
  WSNs, 172–73
Node density, routing protocol comparison, 44–45
Normalized route load (NRL)
  IEEE 802.15.4 network scenarios, 207–8
  number of sensor nodes versus, 208
  number of sources versus, 207
  patrimonial ZigBee routing protocols, 223–24, 228–29
  simulation time versus, 202
  traffic load versus, 207
NRL Sensorsim, 166
NS-2, 166
  defined, 60
  extensibility, 61
  split-programming model, 60–61

On-demand multicast routing protocol (ODMRP), 25, 33
Operating system (OS)
  MANTIS, 139
  Nano-Qplus, 139
  sensor networks, 138–39
  TinyOS, 139
Optimized link state routing protocol (OLSR), 42, 58–59
  average energy consumed versus mobility speed, 72, 73
  average energy consumed versus traffic load, 82, 83
  average network delay versus traffic load, 77, 78
  average network delay versus with mobility speed, 66, 67
  defined, 58
  network throughput versus mobility speed, 68, 69
  network throughput versus traffic load, 79, 80
  optimization in, 58–59
  PDR versus mobility speed, 64, 65
  PDR versus traffic load, 76, 77
  performance analysis, 43
  routing overhead versus mobility speed, 70, 71
  routing overhead versus traffic load, 80, 81
  transmission of packets with MPR, 58
Organization, this book, 14–16
Overhead analysis
  with clustering mechanisms, 116–20
  in hierarchical routing scheme, 114–15
  introduction to, 111–13
Overhead minimization
  for ad hoc networks connected in Internet, 121–23
  with clustering mechanisms, 116–20
  by header compression, 120–21
  introduction to, 111–13
Overlap, 156

Packet delivery ratio
  community model, 100
  defined, 61
  evaluation of, 63–65, 76
  GFMM, 100
  IEEE 802.15.4 network scenarios, 202–4
  with mobility speed, 63–65
  number of nodes versus, 73, 104
  patrimonial ZigBee routing protocols, 218–19, 226
  pause time versus, 105
  to SINK versus number of sensor nodes, 204
  to SINK versus number of sources, 203
  to SINK versus simulation time, 201
  to SINK versus traffic load, 203
  with traffic load, 76
  WSNs, 167–69

Pareto on-off traffic, 215
Patrimonial ZigBee routing protocols
    average energy consumed versus traffic load, 224–25
    average network delay versus number of sources, 227
    average network delay versus traffic load, 220–21, 227
    defined, 214
    introduction to, 211–12
    literature background, 212–14
    network throughput versus number of sources, 228
    network throughput versus traffic load, 221–22, 228
    NRL versus number of sources, 229
    NRL versus traffic load, 223–24, 228
    PDR versus number of sources, 226
    PDR versus traffic load, 218–19, 226
    performance evaluation and traffic load effect, 211–29
    results analysis, 218
    simulation environment, 216–17
    simulation parameters, 217
    traffic generators, 214–15
    traffic model, 216
    types of, 214
Performance metrics
    average energy consumed, 62
    average network delay, 61
    energy models, 125–30
    evaluation with traffic load, 73–83
    IEEE 802.15.4, 198–208
    MANETs, 63–83, 100
    mobility models, 100
    packet delivery ratio, 61
    patrimonial ZigBee routing protocols, 218–28
    routing overhead, 62
    throughput, 61–62
    WSNs, 167–73
Poisson on-off traffic, 215
POWER, 144
Priority nodes, 213
Priority traffic, 213
Proactive routing protocols, 150
Proactive-to-proactive routing scheme, 117
Programming models, sensor networks, 160–61
Protocol control overhead (PCO)
    number of actor nodes versus, 171
    performance evaluation, 170–72
    simulation time versus, 171
Pump slowly, fetch quickly (PSFQ), 153

Quality of service (QoS)
    defined, 162
    issues, 163
    routing algorithm, 163
    SAR support, 164
    support, 238

Random direction mobility model, 43
Random walk with reflection (RW-R) mobility model, 87
    billiard reflection function, 92
    defined, 92
Random walk with wrapping (RW-W) mobility model,, 87
    defined, 93
    wrapping function, 93
Random waypoint (RWP) mobility model, 43, 87
    defined, 90
    energy consumption, 129
    phases, 90
Random waypoint-steady state (RWP-SS) mobility model, 87, 88, 90–92
    average network delay, 100
    defined, 90
    energy consumed, 103
    routing overhead, 102
    with/without pause, 92
Rayleigh group force model, 95
Reactive-to-proactive routing scheme, 118
Reactive-to-reactive routing protocol, 117
REALMOBGEN software, 213–14
Received signal strength (RSS), 146
Reference point group mobility (RPGM) model, 87
    checkpoints, 94
    defined, 94
Request zone, 56
Research areas, 239–40
Reverse ad hoc on demand distance vector routing algorithm (RAODV), 49–50
    AODV comparison, 212–13
    defined, 49
Route discovery, 119
Route maintenance, 119

Routing overhead, 116
  defined, 62
  evaluation of, 69–71, 79–81
  group force mobility model (GFMM), 102
  hierarchical routing protocol reduction, 115
  with mobility speed, 69–71
  number of nodes versus, 75, 106
  pause time versus, 107
  RWP-SS, 102
  RW-W, 105
  scalability analysis, 119–20
  with traffic load, 79–81
Routing protocols, 30–31
  adaptive, 150
  classification of, 47
  desired properties in, 45–46
  directed diffusion, 151
  discussion of, 46–60
  literature background, 43–45
  low energy adaptive cluster hierarchy (LEACH), 150–51
  mobility models simulation, 41
  multipath, 150
  multipath design technique, 149
  negotiation-based, 150
  patrimonial ZigBee, 211–29
  proactive, 30, 150
  scenario based performance analysis, 41–83
  sleep mode, 46
  unidirectional/bidirectional links, 46
  in WSNs, 149–50
  See also Specific protocols

Search and rescue operations applications, 236
Secured AODV (SAODV), 25
Security
  authentication, 164
  challenges, 242
  confidentiality, 164
  in MANETs, 35–36
  WSNs, 164–65
Semi-Markov smooth (SMS) mobility model, 98–99
Sensor-MAC (S-MAC), 142, 143
Sensor nodes
  defined, 11
  in wireless sensor networks, 13
Sensor protocol for information via negotiation (SPIN), 154
Sensor query template, 158–59
Sequential assignment routing (SAR) protocol, 164
Simulation environment
  IEEE 802.15.4, 196–97
  MANETs, 61–63
  mobility models, 99–100
  patrimonial ZigBee routing protocols, 216–17
  WSNs, 166–67
Simulations
  environment, 61–63
  mobility models, 99–100
  with NS-2, 61
  parameters, 63
Sleep mode, 46
Slotted CSMA-CA, 188
  backoff periods, 189
  beacon enabled, 189–92
  illustrated, 190
Smart dust, 136
Software
  POWER, 144
  REALMOBGEN, 213–14
  WSN, 138–39
Source routing based multicast protocol (SRMP), 33
Superframe structure
  in beacon-enabled mode, 187
  IEEE 802.15.4, 187–88
Synchronization, 146–47
  challenges, 146–47
  defined, 146
  global clock, 147
  IEEE 802.15.4, 194
  physical ordering, 147
  protocols, 147
  relative notion of clock, 147

Temporally ordered routing algorithm (TORA), 42, 59–60
  average energy consumed versus mobility speed, 71
  average network delay versus with mobility speed, 65
  defined, 59
  directed paths, 60

energy consumption, 114, 129–30
network establishment steps, 59
network throughput versus mobility speed, 67
number of nodes, 127
path establishment messages, 59
PDR versus mobility speed, 65
reverse link, 60
routing overhead versus mobility speed, 69
Time difference of arrival (TDoA), 146
Time of arrival (ToA), 146
TinyDB, 159–60
TinyOS, 139
Topology control, 29–30
in ad hoc networks, 30
interference and, 30
mismanagement of, 29
as multihop network requirement, 29
Traffic load
average energy consumed versus, 224–25
average energy consumed with, 81–83
average network delay versus, 199, 220–21
average network delay with, 76–78
dropped packets versus, 200
IEEE 802.15.4 metric evaluation with, 198–200
network throughput versus, 199, 221–22
network throughput with, 78–79
NRL versus, 223–24
PDR versus, 218–19
PDR with, 76
performance metrics evaluation with, 73–83
routing overhead with, 79–81
Traffic management/monitoring applications, 236
Transport layer, 152–53

Unslotted CSMA-CA, 188, 191

Vaddina energy model, 112, 123
energy exhausted versus transmission range, 126
values, 124

WiMAX, 3–5
applications, 10–11
bandwidth and range, 10
defined, 9
design, 9–10
fixed scenario, 10
IEEE 802.16 standard, 9
as replacement technology, 11
transmitter clusters, 10
Wireless in local loop (WLL), 3
Wireless local area networks (WLANs), 3
Wireless metropolitan area networks (WMANs), 3–5
Wireless networks
benefits of, 1–2
categories, 4
overview of, 1–3
setup, 2
types of, 3–5
wired networks versus, 2–3
WLAN, 3
WMAN, 3–5
WPAN, 3
WWAN, 5
See also Wireless sensor networks (WSNs)
Wireless personal area networks (WPANs), 3
Wireless routing protocol (WRP), 24
Wireless sensor and actor network (WSAN)
coordination, 165–66
defined, 165
Wireless sensor networks (WSNs), 11–13
active research areas, 239–40
analytical and practical result challenges, 243–44
applications, 135–36, 233–37
attacks in, 166
average end-to-end delay based performance evaluation, 169–70
calibration, 147–48
challenges and future scope, 241–44
communication issues, 139–40
control packet overhead based performance evaluation, 170–72
data aggregation and dissemination, 154–56
database centric and querying, 156–60
defined, 11, 133
deployment, 144–45
distributed sensing, 140
energy efficiency, 149
hardware cost challenges, 242

Wireless sensor networks (continued)
  hardware issues, 136–38
  introduction to, 133–34
  literature background, 135
  localization, 145–46
  long-range communication, 140
  low power consumption, 139
  MAC protocols, 141–43
  middleware, 161–62
  multihop networking, 140
  narrowband communication methods, 140
  network layer issues, 148–52
  nodes, 136
  operating system (OS), 138–39
  packet delivery ratio based performance evaluation, 167–69
  popularity of, 12
  popular routing protocols, 150–52
  power challenges, 242
  primary issues for, 237–38
  programming models, 160–61
  quality of service (QoS), 162–64
  real-world protocol challenges, 242
  recent advances and future trends, 244
  recent developments, 238–39
  research challenges, 242–44
  result analysis, 167–73
  routing protocol classification, 149–50
  security, 164–65
  security challenges, 242
  sensor nodes, 11, 13
  simulation environment, 166–67
  software issues, 136–38
  synchronization, 146–47
  system architecture challenges, 242
  throughput based performance evaluation, 172–73
  topology control, 240
  transport layer issues, 152–53
  WSAN, 165–66

Recent Titles in the Artech House Mobile Communications Series

John Walker, Series Editor

3G CDMA2000 Wireless System Engineering, Samuel C. Yang

3G Multimedia Network Services, Accounting, and User Profiles, Freddy Ghys, Marcel Mampaey, Michel Smouts, and Arto Vaaraniemi

802.11 WLANs and IP Networking: Security, QoS, and Mobility, Anand R. Prasad, Neeli R. Prasad

Achieving Interoperability in Critical IT and Communications Systems, Robert I. Desourdis, Peter J. Rosamilia, Christopher P. Jacobson, James E. Sinclair, and James R. McClure

Advances in 3G Enhanced Technologies for Wireless Communications, Jiangzhou Wang and Tung-Sang Ng, editors

Advances in Mobile Information Systems, John Walker, editor

Advances in Mobile Radio Access Networks, Y. Jay Guo

Applied Satellite Navigation Using GPS, GALILEO, and Augmentation Systems, Ramjee Prasad and Marina Ruggieri

Artificial Intelligence in Wireless Communications, Thomas W. Rondeau and Charles W. Bostian

Broadband Wireless Access and Local Network: Mobile WiMax and WiFi, Byeong Gi Lee and Sunghyun Choi

CDMA for Wireless Personal Communications, Ramjee Prasad

CDMA Mobile Radio Design, John B. Groe and Lawrence E. Larson

CDMA RF System Engineering, Samuel C. Yang

CDMA Systems Capacity Engineering, Kiseon Kim and Insoo Koo

CDMA Systems Engineering Handbook, Jhong S. Lee and Leonard E. Miller

Cell Planning for Wireless Communications, Manuel F. Cátedra and Jesús Pérez-Arriaga

Cellular Communications: Worldwide Market Development, Garry A. Garrard

Cellular Mobile Systems Engineering, Saleh Faruque

The Complete Wireless Communications Professional: A Guide for Engineers and Managers, William Webb

EDGE for Mobile Internet, Emmanuel Seurre, Patrick Savelli, and Pierre-Jean Pietri

Emerging Public Safety Wireless Communication Systems, Robert I. Desourdis, Jr., et al.

The Future of Wireless Communications, William Webb

Geographic Information Systems Demystified, Stephen R. Galati

GPRS for Mobile Internet, Emmanuel Seurre, Patrick Savelli, and Pierre-Jean Pietri

GPRS: Gateway to Third Generation Mobile Networks, Gunnar Heine and Holger Sagkob

GSM and Personal Communications Handbook, Siegmund M. Redl, Matthias K. Weber, and Malcolm W. Oliphant

GSM Networks: Protocols, Terminology, and Implementation, Gunnar Heine

GSM System Engineering, Asha Mehrotra

Handbook of Land-Mobile Radio System Coverage, Garry C. Hess

Handbook of Mobile Radio Networks, Sami Tabbane

High-Speed Wireless ATM and LANs, Benny Bing

Interference Analysis and Reduction for Wireless Systems, Peter Stavroulakis

Introduction to 3G Mobile Communications, Second Edition, Juha Korhonen

Introduction to Communication Systems Simulation, Maurice Schiff

Introduction to Digital Professional Mobile Radio, Hans-Peter A. Ketterling

Introduction to GPS: The Global Positioning System, Ahmed El-Rabbany

An Introduction to GSM, Siegmund M. Redl, Matthias K. Weber, and Malcolm W. Oliphant

Introduction to Mobile Communications Engineering, José M. Hernando and F. Pérez-Fontán

Introduction to Radio Propagation for Fixed and Mobile Communications, John Doble

Introduction to Wireless Local Loop, Broadband and Narrowband, Systems, Second Edition, William Webb

IS-136 TDMA Technology, Economics, and Services, Lawrence Harte, Adrian Smith, and Charles A. Jacobs

Location Management and Routing in Mobile Wireless Networks, Amitava Mukherjee, Somprakash Bandyopadhyay, and Debashis Saha

LTE Air Interface Protocols, Mohammad T. Kawser

Mobile Data Communications Systems, Peter Wong and David Britland

Mobile IP Technology for M-Business, Mark Norris

Mobile Satellite Communications, Shingo Ohmori, Hiromitsu Wakana, and Seiichiro Kawase

Mobile Telecommunications Standards: GSM, UMTS, TETRA, and ERMES, Rudi Bekkers

Mobile Telecommunications: Standards, Regulation, and Applications, Rudi Bekkers and Jan Smits

Multiantenna Digital Radio Transmission, Massimiliano Martone

Multiantenna Wireless Communications Systems, Sergio Barbarossa

Multi-Gigabit Microwave and Millimeter-Wave Wireless Communications, Jonathan Wells

Multipath Phenomena in Cellular Networks, Nathan Blaunstein and Jørgen Bach Andersen

Multiuser Detection in CDMA Mobile Terminals, Piero Castoldi

OFDMA for Broadband Wireless Access, Slawomir Pietrzyk

Personal Wireless Communication with DECT and PWT, John Phillips and Gerard Mac Namee

Practical Wireless Data Modem Design, Jonathon Y. C. Cheah

Prime Codes with Applications to CDMA Optical and Wireless Networks, Guu-Chang Yang and Wing C. Kwong

Quantitative Analysis of Cognitive Radio and Network Performance, Preston Marshall

QoS in Integrated 3G Networks, Robert Lloyd-Evans

Radio Engineering for Wireless Communication and Sensor Applications, Antti V. Räisänen and Arto Lehto

Radio Propagation in Cellular Networks, Nathan Blaunstein

Radio Resource Management for Wireless Networks, Jens Zander and Seong-Lyun Kim

Radiowave Propagation and Antennas for Personal Communications, Third Edition, Kazimierz Siwiak and Yasaman Bahreini

RDS: The Radio Data System, Dietmar Kopitz and Bev Marks

Resource Allocation in Hierarchical Cellular Systems, Lauro Ortigoza-Guerrero and A. Hamid Aghvami

RF and Baseband Techniques for Software-Defined Radio, Peter B. Kenington

RF and Microwave Circuit Design for Wireless Communications, Lawrence E. Larson, editor

Sample Rate Conversion in Software Configurable Radios, Tim Hentschel

Signal Processing Applications in CDMA Communications, Hui Liu

Smart Antenna Engineering, Ahmed El Zooghby

Software Defined Radio for 3G, Paul Burns

Spread Spectrum CDMA Systems for Wireless Communications, Savo G. Glisic and Branka Vucetic

Technologies and Systems for Access and Transport Networks, Jan A. Audestad

Third Generation Wireless Systems, Volume 1: Post-Shannon Signal Architectures, George M. Calhoun

Traffic Analysis and Design of Wireless IP Networks, Toni Janevski

Transmission Systems Design Handbook for Wireless Networks, Harvey Lehpamer

UMTS and Mobile Computing, Alexander Joseph Huber and Josef Franz Huber

Understanding Cellular Radio, William Webb

Understanding Digital PCS: The TDMA Standard, Cameron Kelly Coursey

Understanding GPS: Principles and Applications, Second Edtion, Elliott D. Kaplan and Christopher J. Hegarty, editors

Understanding WAP: Wireless Applications, Devices, and Services, Marcel van der Heijden and Marcus Taylor, editors

Universal Wireless Personal Communications, Ramjee Prasad

WCDMA: Towards IP Mobility and Mobile Internet, Tero Ojanperä and Ramjee Prasad, editors

Wireless Communications in Developing Countries: Cellular and Satellite Systems, Rachael E. Schwartz

Wireless Communications Evolution to 3G and Beyond, Saad Z. Asif

Wireless Intelligent Networking, Gerry Christensen, Paul G. Florack, and Robert Duncan

Wireless LAN Standards and Applications, Asunción Santamaría and Francisco J. López-Hernández, editors

Wireless Sensor and Ad Hoc Networks Under Diversified Network Scenarios, Subir Kumar Sarkar

Wireless Technician's Handbook, Second Edition, Andrew Miceli

For further information on these and other Artech House titles, including previously considered out-of-print books now available through our In-Print-Forever® (IPF®) program, contact:

Artech House
685 Canton Street
Norwood, MA 02062
Phone: 781-769-9750
Fax: 781-769-6334
e-mail: artech@artechhouse.com

Artech House
16 Sussex Street
London SW1V 4RW UK
Phone: +44 (0)20 7596-8750
Fax: +44 (0)20 7630-0166
e-mail: artech-uk@artechhouse.com

Find us on the World Wide Web at: www.artechhouse.com